BestMasters

Springer awards „BestMasters" to the best master's theses which have been completed at renowned universities in Germany, Austria, and Switzerland.

The studies received highest marks and were recommended for publication by supervisors. They address current issues from various fields of research in natural sciences, psychology, technology, and economics.

The series addresses practitioners as well as scientists and, in particular, offers guidance for early stage researchers.

Roman Feldbauer

Machine Learning for Microbial Phenotype Prediction

Roman Feldbauer
Wien, Österreich

BestMasters
ISBN 978-3-658-14318-3   ISBN 978-3-658-14319-0 (eBook)
DOI 10.1007/978-3-658-14319-0

Library of Congress Control Number: 2016940340

Springer Spektrum
© Springer Fachmedien Wiesbaden 2016

Printed on acid-free paper

This Springer Spektrum imprint is published by Springer Nature
The registered company is Springer Fachmedien Wiesbaden GmbH

Kurzfassung

Genomdatenbanken wachsen rasant. Moderne Metagenomikstudien führen zu einer großen Anzahl annähernd vollständiger Genomsequenzen nicht kultivierbarer mikrobieller Spezies. Diese Entwicklungen führen zur Notwendigkeit der Entwicklung automatisierter bioinformatischer Methoden für die Vorhersage mikrobieller Phänotypen, um die biologische und ökologische Interpretation der großen Datenmengen zu ermöglichen.

In dieser Arbeit wird untersucht, wie komparative Genomik für diesen Zweck eingesetzt werden kann. Verschiedene bioinformatische Prototypen sowie Techniken des maschinellen Lernens werden verglichen. Im Fokus stehen dabei große Genomdatenbanken und inkomplette Genomsequenzen. Darüberhinaus werden notwendige Verbesserungen an der Software vorgenommen. Ein Programm wurde in der Evaluationsphase ausgewählt. Die Stabilität der Vorhersagen phänotypischer Charakteristika wurde im Lichte schnell wachsender Genomdatenbanken demonstriert. Ein neu entwickeltes Softwarewerkzeug ermöglicht die eingehende Analyse von Phänotypmodellen und assoziierte erwartete sowie unerwartete Proteinfunktionen mit bestimmten Merkmalen. Ein Großteil der Merkmale konnte zuverlässig in lediglich zu 60-70% kompletten Genomen vorhergesagt werden. Hochakkurate Modelle wurden für die Vorhersage zweier ökologisch relevanter metabolischer Merkmale (Methanotrophe und Nitrifikanten) erstellt. Sie fanden bereits bekannte funktionelle Marker und erweiterten das Markerkonzept durch die Assoziation weiterer Gene zu den Phänotypen substantiell. Darüber hinaus wurde ein Phänotypmodell für die Vorhersage intrazellulärer Mikroorganismen etabliert. Damit konnte gezeigt werden, dass auch unabhängig evolvierte Merkmale, die durch Genomreduktion charakterisiert sind, zuverlässig durch komparative Genomik vorhergesagt werden können. Alle Modelle wurden mit den Daten aus drei unterschiedlichen Metagenomen getestet. Sie sagten Merkmale voraus, die in Einklang mit den vorherrschenden Umweltbedingungen stehen.

Die Ergebnisse legen nahe, dass die automatische Annotation von Phänotypen in annähernd kompletten mikrobiellen Genomen möglich ist.

Abstract

Public genome databases grow rapidly due to improvements in high-throughput sequencing technology. A large number of almost complete genome sequences of uncultivable microbial species arise from modern metagenomic studies. These developments lead to the necessity of developing automated bioinformatic methods for microbial phenotype prediction, which are solely based on genomic sequences, to enable biological and ecological interpretation of these data.

This thesis is concerned with the investigation of how comparative genomics can be utilized for the prediction of microbial phenotypes. Different prototypic bioinformatic tools and machine learning techniques for phenotypic trait prediction are compared with focus on applicability to large-scale genome databases and incomplete genome sequences. Software improvements are introduced, where necessary. A software tool was selected in the evaluation phase. The stability of its predictive power for phenotypic traits not perturbed by the rapid growth of genome databases was demonstrated. A newly developed program facilitates the in-depth analysis of phenotype models, which associate expected and unexpected protein functions with particular traits. Most of the traits could be reliably predicted in only 60-70% complete genomes. Highly accurate models were created for the prediction of methanotrophs and nitrifiers, two ecologically important metabolic traits. They recovered known functional markers and substantially extended the marker concept by associating further genes to the phenotypic traits. In addition, a new phenotypic model that predicts intracellular microorganisms was established. Thereby it could be demonstrated that also independently evolved phenotypic traits, characterized by genome reduction, can be reliably predicted based on comparative genomics. All models were tested on data from three different metagenomes. They predicted phenotypes that were in alignment with environmental conditions.

The results suggest that the improved prediction tool can be used to automatically annotate phenotypes in near-complete microbial genome sequences, as generated in large numbers in current metagenomics studies.

Contents

List of Figures   XI

List of Tables   XIII

1   Introduction   **1**

   1.1   Biological Background 3

      1.1.1   Genotype and Phenotype 3

      1.1.2   Microbial Genotypes and Phenotypes 6

      1.1.3   Alternative Approaches for Genotype-Phenotype Relations 7

   1.2   Computational Background 9

      1.2.1   Basics of Machine Learning 9

      1.2.2   Association Rule Mining 14

      1.2.3   Support Vector Machines 16

      1.2.4   Databases for Orthologs 22

2   Materials and Methods   **25**

   2.1   Genotype data 25

   2.2   Phenotype data 27

   2.3   Accuracy Measures 27

   2.4   KRONA 29

   2.5   NetCAR 29

   2.6   PICA 32

   2.7   GenTraitor 38

3 Results and Discussion   **41**

   3.1 Comparison and Evaluation of Phenotype Prediction Packages 41

       3.1.1 NetCAR 41

       3.1.2 PICA 44

       3.1.3 Complexity Estimation 47

       3.1.4 Prediction on Incomplete Genomes 50

   3.2 Extension to Novel Phenotypes 54

       3.2.1 Methane Oxidation 54

       3.2.2 Nitrification 58

       3.2.3 Intracellular lifestyle 62

   3.3 Phenotype Prediction in Metagenomics 67

       3.3.1 Coral metagenome 69

       3.3.2 Sponge metagenome 70

       3.3.3 Biogas-fermenter metagenome 71

4 Conclusion and Outlook   **79**

Abbreviations   **83**

References   **85**

A Source code   **97**

B Tables   **105**

C Acknowledgments   **109**

List of Figures

1.1 Model selection 13

1.2 Sparse kernel machine 18

1.3 Maximum margin classifier 19

1.4 Soft margin support vector machine (SVM) 20

1.5 XOR function 20

2.1 Accuracy, balanced accuracy, F1-score 29

2.2 Conditional Mutual Information 34

3.1 CPAR function check 44

3.2 ARM prediction quality 45

3.3 ARM runtime 46

3.4 ARM vs. SVM prediction quality 47

3.5 ARM vs. SVM runtime 48

3.6 SVM kernel selection 49

3.7 eggNOG 2.0 vs. eggNOG 4.0 runtime 50

3.8 eggNOG 2.0 vs. eggNOG 4.0 prediction quality 51

3.9 NOG vs. bactNOG prediction quality 52

3.10 Scaling database sizes and problem dimensionality 53

3.11 Phenotype prediction for incomplete genome sequences 53

3.12 Taxonomy of methanotrophs in the training set 55

3.13 Taxonomy of non-methanotrophs in the training set 55

3.14 Prediction of methanotrophs in eggNOG 4.0 58

3.15 Hierarchy of nitrification models 58

3.16 Taxonomy of nitrifiers in the training set 59

3.17 Taxonomy of non-nitrifiers in the training set 59

3.18 Predicted nitrifiers in eggNOG 4.0 62

3.19 Taxonomy of intracellular bacteria in the training set 63

3.20 Taxonomy of free living bacteria in the training set 64

3.21 Taxonomy of predicted obligate intracellular bacteria in eggNOG 4.0 65

3.22 Black band disease and cyanobacterial patches 69

List of Tables

3.1   Number of distinct k-combinations of 11 969 COGs 42

3.2   NetCAR: number of association rules vs. number of unique COGs 42

3.3   NetCAR: best rules for five phenotypes 43

3.4   Methanotroph model feature ranking 57

3.5   Nitrification models accuracy 60

3.6   Predicted obligate intracellular archaea in eggNOG 4.0 67

3.7   Predicted obligate intracellular microbial eukaryotes in eggNOG 4.0 67

B.1   Nitrification model feature ranking 106

B.2   Obligate intracellular model feature ranking 108

Chapter 1

Introduction

Microorganisms, comprising bacteria, archaea and unicellular eukaryotes, are key components of all ecosystems on earth. Their tremendous phylogenetic, ecological and functional diversity is so far only insufficiently understood. Although genome sequencing has within the last two decades enormously advanced the investigation of microbes, microbial genomes have mainly been sequenced from DNA obtained from well-characterized, pure lab cultures. The majority of microbes, however, cannot be cultivated and was therefore inaccessible for genome research [1]. Metagenomic techniques, studying DNA directly obtained from environmental samples, have provided first important insights into genomic features of the unseen majority of microorganisms (recently reviewed in [2]). Improvements in high-throughput sequencing and DNA extraction protocols, combined with advanced computational methods for binning and taxonomic classification of metagenomic sequences, have recently enabled the reconstruction of near-complete genome sequences of even low-abundant members of microbial communities [3][4][5]. These recent advances have not only triggered a paradigm shift from "gene-oriented" to "genome-oriented" metagenomics, but also leave us with an emerging bioinformatic problem: the prediction of biological phenotypes and ecological roles of uncharacterized microbial species from their partial genome sequences.

The representation of microbial genomes by their protein-coding genes, associated to orthologous or homologous groups, is the most widely used approach for the organization of large-scale genomic data [6][7][8]. A wide range of applications, e.g. the prediction of metabolic functions [9] or protein-protein interactions [10], utilize clusters of orthologous groups. A Cluster of Orthologous Groups (COG) is a phylogenetic abstractions of genes on the last universal common ancestor (LUCA) level. Non-supervised Orthologous Groups (NOG) extend this concept to non-supervised groups [7]. So far the highest phylogenetic and ecological diversity of published genome sequences has been achieved for bacteria. During the last years also an increasing number of archaea and unicellular eukaryotes have been included in genome databases [11]. According to fundamental principles of microbial genome evolution, such as the preference for compact and streamlined genomes [12], the presence or absence of COGs in microbial genomes is highly informative. The continuing growth of genome databases therefore holds an enormous potential for advanced computational methods making use of large-scale comparative genomics.

Phenotypic traits of microbes can be very diverse. Structured and computer-readable organization of trait descriptions has been suggested e.g. in the Ontology of Microbial Phenotypes [13]. They range from morphologic and physiological traits to specific molecular or metabolic capabilities. Numerous traits, such as cell envelope structure, as indicated by the Gram stain, have been acquired early in evolution and are therefore encoded in the core section of the pan-genome [14]. Other traits, such as protein secretion capabilities, are evolutionarily more dynamic and are encoded in the variable section of the pan-genome. It can be speculated that the broad evolutionary diversity of microbial traits will be a substantial challenge for computational methods. Generic computational methods, namely those based on large databases of COGs, will most likely represent a first layer of methods for trait prediction. Additional specific models describing well-defined traits based on metabolic and/or regulatory models (e.g. used in [15]) will be needed for a deeper interpretation of a microbial genotype.

Previous work on such generic methods was in many cases limited to searching for one-to-one relations between genes and phenotypic traits. This strategy works very reasonably for simple metabolic traits. E.g., the *amoA* gene encoding ammonia monooxygenase [16] is characteristic for ammonia oxidizing archaea and bacteria. However, single marker genes are of limited predictive power for many other traits. Recent methodological improvements utilize heuristic association rule mining (ARM) to find many-to-one relations. They are based on mutual information [17] or predictive associations [18]. MacDonald and Beiko [18] developed a software framework for comparison of different phenotype prediction methods, Phenotype Investigation with Classification Algorithms (PICA), which operates on the level of COG presence or absence in genomes. It includes plug-ins for classification based on predictive association rules (CPAR) and libSVM [19] among others. While PICA features support vector classification (SVC) with the latter, the original authors focused on ARM and a conditionally weighted mutual information (CWMI) metric.

In this thesis I investigate the application of different machine learning techniques for phenotypic trait prediction in microbiology and microbial ecology, with an emphasis on metagenomic sequences and vastly increasing data amounts. I further present novel models for prediction of methanotrophic, nitrifying or intracellular microorganisms. All models are finally used for phenotype prediction of genomic bins in three different metagenomes.

1.1 Biological Background

1.1.1 Genotype and Phenotype

General definitions and examples

The concept of genotype and phenotype is one at the very base of biology. Campbell biology defines the genotype as "[t]he genetic makeup, or set of alleles, of an organism"[20, G-15] and the phenotype as "[t]he observable physical and physiological traits of an organism, which are determined by its genetic makeup."[20, G-26].

In classical Mendelian genetics, the phenotype is determined by the combination of alleles (alternative versions of the same gene). Alleles are usually either dominant or recessive, where the trait encoded by the recessive allele is only observed, if no dominant alleles are present. Organisms with diploid cells have two alleles (one on each chromosome). Consider a diploid plant with either red or white flowers, where the R and r alleles code for red and white, respectively, and R is dominant over r. Then two different genotypes (RR homozygous and Rr heterozygous) exhibit the red trait, whereas only the rr homozygous genotype causes white flowers. Thus, the distinction between genotype and phenotype.

Other definitions describe the genotype is a broader sense: It may also refer to "an individual's collection of genes"[1]. Different phenotypic traits can on the one hand be caused by minor variations on the DNA sequence level, like e.g., insertions and deletions (indels), which are frameshift mutations that lead to dramatic changes in the protein sequence. For example, susceptibility to Crohn's disease is associated with an insertion in the *NOD2* gene[21] and inherited resistance to HIV-1 is based on a frameshift mutation in *CCR5* [22]. On the other hand, the very presence or absence of a gene can cause a certain trait. This is, for instance, the case for many metabolic traits in bacteria that rely on certain enzymes: Nitrifiers require an ammonia monooxygenase (AMO, compare Section 3.2.2), methanotrophs oxidize methane via the methane monooxygenase (*MMO*, compare Section 3.2.1), while the dissimilatory (bi)sulfite (*dsrAB*) and adenosine-5'-phosphosulfate reductases (*apsA*) are key enzymes for sulfate-reducing prokaryotes [23]. Thus, a suitable representation of the genetic makeup is required for inferring phenotype from genotype.

The complexity of genotype-phenotype relations

The case of one genetic feature with two manifestations causing two different traits as described above is the simplest form of a genotype-phenotype relation. Besides polyploidy, several other factors exacerbate the problem of finding such relations [20]. Multiple allelic forms may exist for every gene. One well-known example are the AB0 blood groups in humans, where A and B are co-dominant over 0.

Another factor is pleiotropy, which is a gene's property of influencing two or more phenotypic traits. In complex organs such as the human brain pleiotropic effects are widespread: For example, the peptide Oxytocin is known to the public as the "cuddling hormone" for its role in sexual reproduction and pair bonding. However, Oxytocin, which acts on inhibitory interneurons, has recently been shown to regulate a large number of systems, like facial recognition, empathy, social cognition and might even play a role in a subtype of autism [24].

While one gene may affect several phenotypic traits, the reverse is possible as well: Epistasis is the effect of functional gene interaction, where one gene's phenotypic expression changes that of another gene at an independently inherited locus [20].

---

[1]http://www.genome.gov/glossary/index.cfm?id=93, National Human Genome Research Institute, NIH, USA. Retrieved on 2015/08/28.

Polygenic inheritance describes phenotypic traits that are caused by additive effects of two or more genes. These traits often are non-binary, but quantitative characters, like for instance human body height with at least 180 associated genes [20] or skin pigmentation. Another example is the Type III secretion system, a pathogenicity factor in gram-negative bacteria. These structures are built from roughly 30 different proteins, some of which are shared with the flagellum, which enables motility. In this case, genotype-phenotype relations indeed have many-to-many cardinality. Predictive models thus need to capture this complexity.

The genotype is not enough or: The influence of regulation and the environment

Moreover, only considering an organism's genotype often limits to propositions about its phenotypic potential, because the actual trait is also influenced by the environment. This includes abiotic, biotic, social or even economical and political factors. For example, skin pigmentation changes with exposure to sunlight, constitution is dependent on diet and nutrients (among other factors), and general health status is correlated with social stratification.

All organisms utilize nucleic acids to store their genetic information. This information must be turned into functional products in order to cause a phenotypic trait. Gene expression is the process of transcribing genes into RNAs, which may themselves be functional or are otherwise further translated into proteins. These biosynthetic processes are usually subject to regulatory mechanisms. Especially multicellular organisms make use of complex regulation on all levels of gene expression: In addition to transcriptional and translational regulation, also post-translational modifications alter product activity. Furthermore, eukaryotes organize their genes in exon-intron architectures and have the capability to produce different gene products from one gene through alternative splicing.

Epigenetics is the study of heritable phenotypic traits that are caused by changes to the chromosome without altering the nucleotide sequence: Covalent modifications to histones and the DNA itself play a role in transcriptional regulation. Moreover, cells of higher eukaryotes typically express only small subsets of their genes. These subsets are distinct for different tissues despite identical genomes. Differential gene expression allows for differences between cell types. It is thus clear, that the shear presence of a gene might often be insufficient to infer a phenotypic trait, in particular so for higher eukaryotes.

1.1.2   Microbial Genotypes and Phenotypes

Regulation in bacteria is typically less complex than in eukaryotes. A bacterial cell contains a single circular chromosome, possibly plasmids, no histones, no nucleus and comparatively simple genetic architectures without introns. Thus, many regulatory mechanisms of eukaryotes are irrelevant in bacteria, which, for example, have neither chromatin remodeling nor nucleic export regulation. Since prokaryotes are generally considered haploid, there is usually no need to consider effects due to homozygous versus heterozygous states. Genomes of *Archaea* and *Bacteria* are smaller and more compact than those of *Eukarya* [12]. The number of genes in prokaryotes is typically in the order of one to several thousands [25]. The amount of non-coding DNA is usually very low and there appears to be a linear correlation between genome size and the number of open reading frames (ORFs, sequences that may harbor genes). Sizes of prokaryotic genomes are in the order of few megabase pairs (Mbp), where obligate parasites such as *Nanoarchaeum equitans* can have genomes shorter than 500 kbp.

In comparison, eukaryotic genomes contain large intergenic regions, and protein-coding sequences of genes (exons) may be separated by non-coding sequences (introns). The density and content of introns varies considerably in *Eukarya*: While in unicellular eukaryotes often only a small subset of all genes have exon-intron organization, higher eukaryotes like, e.g., vertebrates on average have several introns per gene [12]. Koonin [12] suggests a coarse categorization of genomes into (1) Small, compact genomes of viruses, prokaryotes and many unicellular eukaryotes, which are predominantly composed of coding sequences (2) Large, expansive genomes of multicellular and particular unicellular eukaryotes, large fractions of which are non-coding sequences. Phenotype inference based on gene presence appears to be justified in the streamlined genomes of the first category. For the second category, however, more detailed models are most likely necessary to predict phenotypes. The scope of this thesis is, therefore, limited to phenotype prediction in microorganisms.

An emphasis will be put on metagenomics, which is the field of genomic studies of uncultured microbes [26]. It is also referred to as environmental genomics, because the microorganisms are sampled from their habitats. Classical microbial genomics requires the (often difficult) cultivation of microbes. Clonal cultures for one facilitate the reconstruction of complete genomes. Moreover, they enable a plethora of other wet-lab experiments to characterize the organism. For example, a simple test for the ability to synthesize tryptophane is to let the microbe grow on a medium lacking this amino acid. The phenotype of a cultured microorganism can hence be examined directly. In contrast, many experimental techniques are not applicable in metagenomic context. The ability to predict phenotypic traits solely based on environmental genomic data could thus greatly enhance our understanding of the corresponding habitats.

Hereinafter, phenotype prediction tools will be evaluated with a special focus on applicability to metagenomic data. Tools will be extended or newly developed, if necessary. New models for the prediction of ecologically important traits will be created and finally applied to genomic as well as metagenomic sequences.

1.1.3 Alternative Approaches for Genotype-Phenotype Relations

Besides the methodology of relating gene presence and absence to phenotypic traits as suggested in this thesis, several other approaches have been developed so far. A selection of those techniques is presented in the section below.

Genome-wide association studies (GWAS) investigate statistical relations between genetic variants (usually single-nucleotide polymorphisms, SNPs) and certain traits. They are most commonly applied in epidemiology to examine genetic underpinnings of human diseases. GWAS was first used in 2005 in a study about age-related macular degeneration, in which two SNPs were found to be associated with the disease [27]. Since then, GWA studies have been used for many diseases, like e.g., diabetes, arthritis or Crohn's disease. Hundreds of loci have been found to be associated with obesity [28]. The proportion of explained variance for the body mass index (BMI) by these variants is, however, low [29]. Fundamental criticism of GWA studies thus concerns the impact of genetic variants on human diseases in relation to environmental and other factors. Expert statements range from praising GWA for revolutionizing our understanding of disease to GWA studies not being worth the expenditure [30].

In some cases, phenotypic traits can directly be inferred from taxonomy or phylogeny. For example, the presence of mammary glands is the basis for the taxonomic clade *Mammalia*.

Any genome phylogenetically classified as mammalian thus corresponds to an organism capable of lactation. In microbiology, cell wall properties as indicated by the Gram stain can largely be deduced from taxonomy, though neither gram-positive nor gram-negative bacteria form monophyletic groups. Still, for instance, any genome found to correspond to a proteobacterium can safely be predicted to bring forth a thin peptidoglycan layer, since all proteobacteria are gram-negative. On the contrary, predictions are harder within *Firmicutes*: Although most species in this phylum have gram-positive cell wall structures with a thick peptidoglycan layer, members of the class *Negativicutes* possess a double lipid bi-layer, causing gram-negative staining [31]. A third example stems from mammalian gut microbiome research. Studies in mice as well as humans comparing the microbiomes of lean and obese individuals found that the relative abundance of *Bacteroidetes* and *Firmicutes* was associated with obesity [32]. Gut microbiota would thus be an additional factor to obesity pathophysiology. Recent studies, however, question this association. Finucane *et al.* report no associations between BMI and stool microbiomes based on the large dataset from the Human Microbiome Project [33].

Biological functions are often carried out by more than a single protein. Prokaryotes often group genes that serve a common task into operons. For example, the *lac* operon of *Escherichia coli* comprises three genes for metabolism and transport of lactose. They are regulated collectively in dependence of lactose and glucose availability. The presence of a *lac* operon clearly indicates the Lac$^+$ phenotype, i.e., the ability to use lactose as carbon and energy source. The individual genes, however, are insufficient for the expression of the phenotype (or its prediction): Both *lacZ*, coding for the enzyme that degrades lactose to monosaccharides, and *lacY*, a transporter for lactose uptake, are required for lactose catabolism. Protein complexes are another type of protein-protein interactions that perform numerous functions in the cell. Examples range from oxygen transport in the blood via haemoglobin or inhibitory neurotransmission in the central nervous system via the GABA$_A$ receptor to transcription via RNA polymerases. Proteins may also interact through co-operative binding or allosteric effects, and also transport, sequestering as well as signaling interactions occur [10]. Databases, such as STRING, aim for association of proteins into functional sets or modules. These functional associations may then allow for insights about the module's biological purpose.

1.2 Computational Background

1.2.1 Basics of Machine Learning

The concept of learning

The Encyclopædia Britannica defines the process of learning as "the alteration of behavior as a result of individual experience". An organism's prerequisites for learning are its ability to perceive and change its behavior [34]. The definition is very broad in order to include a large variety of different learning types that range from classical conditioning and habituation to imitation and problem solving etc. In living organisms, these processes are dependent on the species' central and peripheral nervous systems. Different learning types often require highly specialized brain areas for proper functioning. For example, concept acquisition in humans necessitates a neural circuit containing the hippocampus and the ventromedial prefrontal cortex [35]. Thus, species with different levels of brain complexity also differ in the range of learning types they can employ, which is strongly linked to their potential intelligence.

Extending the concept to artificial systems

The concept of learning is not limited to biological organisms. Artificial systems can be designed in a way, that allows for modification of actions based on experience. Machine learning is thus considered a subfield of artificial intelligence research. It requires algorithms, which improve performance through experience. In order to achieve this, such algorithms need to perceive their "behavior" and the corresponding output, e.g., success in a given task or the result of a defined loss function. They are usually tailored to specific, encapsulated tasks, with some canonical examples ranging from spam filtering and visual pattern recognition to playing chess. Machine learning techniques are being applied to numerous problems in natural sciences as well as in the corporate sector and the industry: Hidden Markov models (HMM) are an example of stochastic models successfully used in many bioinformatic areas, like gene prediction or modeling protein sequence families [36]. HMMs are also used in speech synthesis to model acoustic parameters [37] and in cryptanalysis for modeling randomized side-channel countermeasures [38] or for modeling human-memorable passwords in "smart-dictionary" attacks [39]. Facebook, Inc. in 2014 created a facial recognition software based on deep learning algorithms, which they claim is 97% accurate, thus effectively performing as well as humans in the task [40]. Recent advances in computer science allow robots to quickly adapt to injuries based on a trial-and-error learning algorithm [41], similar to animals in comparable situations. While machine learning is used for very different tasks, there is not one single technique that fits all. Instead, various different techniques have been developed so far. Comparable to different regions of the brain, different techniques have special capabilities and limitations. Indeed, so-called artificial neural networks were originally inspired from biological neural networks. Thus, the field of machine learning is on many levels a highly interdisciplinary one and a strong link between biology and computer science.

Types of machine learning

There are several characteristics to a specific learning scenario. One important aspect is the response an algorithm receives upon producing an output. This relates to the question of how the algorithm can improve in a given task [42, Chapter 1.3]. Depending on the type of feedback different categories arise:

- **Supervised learning** This is a scenario of learning with a supervisor or teacher, who provides training examples with desired outputs (targets, ground truth). Based on a finite number of input data, the learning system has to generalize in order to map all possible inputs to the correct corresponding output. Based on the type of output, one can differentiate between several sub-categories:

    Classification is the task of assigning labels to data points, depending on groups they belong to. For example, scenery image recognition aims to assign photos to classes depending on their content, like airplanes, amusement parks, pagodas, etc. Many classification tasks do not involve multiple classes, but only require binary decisions, like classifying an email as spam or not. The major goal of this thesis - microbial phenotype prediction - will be considered a task of binary classification: the algorithm decides between phenotype presence and absence. In any case, the output in classification tasks is limited to a finite number of discrete values [43, Chapter 1].

If the output can take values on a continuous scale, the problem is called *regression*. Typical examples are the prediction of stock market prices or the determination of planetary orbits from astronomical observations, with the latter dating back to the work of Legendre and Gauß in the early 19th century.

- **Unsupervised learning** No knowledge about the ground truth is given, i.e., there are no labeled training data. The unsupervised learning system is required to find similarities and structures in the input data on its own. A typical task is *clustering*, where inputs are grouped together based on these findings. In contrast to the supervised classification task, no specific classes are defined a priori, although in some cases the number of desired clusters needs to be fixed in advance (e.g. in k-means, where the inputs are divided into k clusters). Other unsupervised tasks are finding the distribution of data in their input space (*density estimation*) or mapping inputs into a lower-dimensional space (*dimensionality reduction*), especially as projections to two or three dimensions in order to enable *visualization* of large data sets [43].

- **Reinforcement learning** In this task somewhere between supervised and unsupervised learning [42], software agents need to find suitable actions in a dynamic environment in order to maximize success. While there are no pairs of inputs with optimal outputs available, a process of trial and error leads to a result, which can be evaluated as favorable or not. For example, consider an algorithm learning to play a game like Backgammon. In a typical course of any game, good moves as well as bad ones are carried out by all players, especially if played in a trial and error fashion. It is often not possible to discern between good and bad moves immediately, but only victory or defeat at the end of a game give certain feedback about how good the sequence of all moves was. The learning system, thus, needs to adequately attribute the reward of success to only the good moves [43]. Marsland [42] compares *reinforcement learning* to learning with a *critic*, who gives grades, but does not give indications, how performance could be increased.

Some authors use additional categories like *semi-supervised* learning (combining training data with and without labels) or *evolutionary learning* (adapting the biological concept of *survival of the fittest*). In the following sections, two machine learning techniques will be highlighted: ARM and SVM. These two methods will then be evaluated with respect to applicability to microbial phenotype prediction.

Model quality assessment and parameter selection

The general process of machine learning contains several phases [44, Chapter 1.5], which shall discussed here in the light of phenotype prediction.

First, data needs to be collected and prepared. This often requires difficult or expensive measurements. Phenotype prediction necessitates complete genomes for training, thus, clonal cultures of microbial organism need to be established, which is still hard for many species. Genomic data can then be obtained by high-throughput sequencing techniques. Only the rapid development of this field makes data acquisition affordable and makes, therefore, phenotype prediction possible. Public resources such as the NCBI Genomes database provide several thousand complete prokaryotic genomes. While getting hold of input data appears to be rather simple today, obtaining target data still poses a major problem. Due to a lack of comprehensive phenotype databases, expert knowledge is often required to assign phenotypic trait labels to genomes.

Second, feature selection is required, i.e., the identification of features that are suitable for the given task. For simple tasks, common sense might suffice. Consider the task of recognizing Euro coins: Size and weight are obviously helpful to distinguish between different coins, whereas shape encodes no useful information. For phenotype prediction, we decided to utilize orthologous groups as features, because from general biological knowledge it is clear that there is a strong link between gene presence and phenotypes (compare Section 1.1.1). Since the plain data are nucleotide sequences, it would be also possible to use GC content, k-mer frequencies, the total number of proteins, the length of intergenic regions or others as features. Their predictive power would be, however, questionable.

Third, an appropriate algorithm must be chosen for the particular problem. This requires knowledge of machine learning principles and of the structure of the data as well as the problem

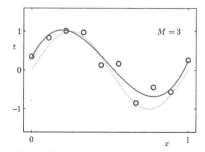 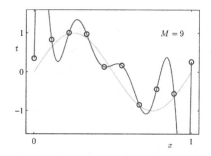

Figure 1.1: Model selection in a curve fitting problem. The light green line indicates the actual process, from which ten datapoints have been generated with some noise (blue circles). The red lines represent two polynomial models fitted to the data. The polynomial of degree three roughly approximates the original function, while the higher degree polynomial is over-fitted. Figure modified after [43, Ch.1.1 Example: Polynomial Curve Fitting, Fig. 1.4, p.7] with permission of Springer.

itself. In phenotype prediction, we want to decide whether an organism exhibits a certain trait or not. We do this by considering other organisms, for which their trait status is known. Thus, supervised learning algorithms for classification are required. Choosing an appropriate one is one part of this thesis (compare Section 3.1.2).

Fourth, the model and its parameters (where applicable) need to be set. Special care has to be taken that the model is sufficiently complex, but not more complex than necessary. Fig. 1.1 illustrates this problem in a simple setting. When using SVC for phenotype prediction, a suitable kernel needs to be selected as well as the soft margin parameter C (see Section 1.2.3) and possibly further parameters. It is often impossible to know good choices for these a priori. Generally, it is necessary to empirically test for good models and parameters. This is done by splitting the data into three parts: The *training set* is used to create a model, which is subsequently confronted with the *validation set* in order to estimate its performance on previously unobserved data. An additional *test set* can be used to avoid overfitting on the validation set. When available data is scarce, it may be desirable to use each datapoint for training. In this case multi-fold cross-validation schemes prove to be useful. In k-fold cross-validation, the data is split randomly into k subsets. One of these subsets is held out as a validation set, while the model is trained on the remaining subsets. The process is repeated

k times, so that each subset is used as validation set exactly once. One cross-validation is performed for each combination of model types and values, which can be a very costly process. The cross-validation with the lowest validation error decides on the model and parameters. The accuracy achieved in cross-validation can be interpreted as a rough estimation of the model's general performance.

The last two phases, training and evaluation, have already been described above. In phenotype prediction using ARM, training is the computational process of finding association rules within the training set that are highly predictive for a certain trait. Subsequently, these rules are evaluated on different data using some cost function or accuracy measure.

1.2.2   Association Rule Mining

ARM is a field of machine learning for detection of dependencies between categorical variables [45]. ARM techniques search for strong rules in large databases $D = \{t_1, \ldots, t_k\}$, containing k transactions. The rules are of the form $X \Rightarrow Y$, where X and Y are sets of features. The left-hand side is also called antecedent, the right-hand side is referred to as consequent. An intuitive interpretation of association rules might read as follows: Whenever X is observed in a transaction, Y is most likely present as well.

Association rule mining in market basket analysis

In general ARM, X and Y are subsets of I, where $I \in \{i_1, \ldots, i_w\}$ is a set of w items. ARM was introduced by Agrawal et al. [46] in the context of market basket analysis. There, I might for example reflect the set of all consumer goods offered by a supermarket. Association rules can be used to describe co-occurrence of certain products within single customer transactions. For example it may occur that transactions containing unscented lotion, food supplements of $Ca^{2+}/Mg^{2+}/Zn^{2+}$ and large amounts of soap often also contain large amounts of cotton balls (adapted from [47]).

ARM is not limited to finding interesting correlations between items of a common set, but can also be applied as a technique of supervised learning for classification. Assume $I = G \cup P$, so that $G \cap P = \{\}$. Then $X \subseteq G = \{g_1, \ldots, g_n\}$, a set of n binary attributes, and

$Y \subseteq P = \{p_1, \ldots, p_m\}$, a set of m binary attributes of another type. Consider an example, where G contains consumer goods as in the example above. Then P could be a set of customer properties, like for instance age, sex, social status, etc. ARM can then be used for inference on the customer's background. A bold rule could have the form lotion + supplements + soap $\Rightarrow$ pregnancy.

Association rule mining for phenotype prediction

In a biological context, similar usage scenarios emerge. Instead of product relations, co-occurrence of genetic features might be analyzed there. A minimal example for association rules of gene presence within genomes of single species is 16S rDNA $\Rightarrow$ 23S rDNA. In this thesis, we focus on ARM-based classification for inference on organismic properties. For microbial phenotype prediction, we define G and P as sets of genetic features and phenotypic traits, respectively. Thus, a typical rule might look like the following: citrate synthase + cytochrome c oxidase + peroxiredoxin $\Rightarrow$ aerobic lifestyle [17].

More specifically, we define G as the eggNOG database (see Section 2.1), i.e., the space of all COGs and highest-level NOGs. On the other hand, we do not restrict P a priori, but allow all sensible microbial traits to be members of P. In case of doubt, we discuss reasonableness per trait individually. Following these definitions, the database D would ideally contain the genomes of all microbial species. In practice, however, this is always limited to (1) the currently sequenced and annotated microbial genomes, which are (2) labeled with respect to the phenotype of interest.

Statistical and algorithmic considerations

In the context of ARM, there are some important measures and concepts, which are useful to detect relevant rules among all possible rules (based on [48]):

$$\text{supp}(X) := \frac{|\{t \text{ in } D \,|\, t \text{ contains } X\}|}{|D|} \tag{1.1}$$

The support of a set X is the probability, that a random transaction t contains all the items of X.

$$\text{conf}(X \Rightarrow Y) := \frac{\text{supp}(X \cup Y)}{\text{supp}(X)} \tag{1.2}$$

The confidence of a rule $X \Rightarrow Y$ is the conditional probability, that the rule holds, given X.

$$\text{lift}(X \Rightarrow Y) := \frac{\text{supp}(X \cup Y)}{\text{supp}(X)\text{supp}(Y)} \tag{1.3}$$

The lift is defined as the ratio of the observed support and the expected support (under the assumption of X, Y independent).

In some cases it might be desirable to broaden the scope of association rules. The standard case of *positive* association rules only correlates feature presence. *Negative* association rules expand the framework to absent features and combinations of present and absent features. Here, we will only consider the simple case of $X \Rightarrow \neg Y$, i.e., a species is predicted not to display a certain phenotype, if a defined set of genes is present in the genome. Artamonova *et al.* [49] mined positive and negative association rules in large annotation databases in order to find annotation errors. Indeed, a biologically trivial rule like nuclear protein $\Rightarrow \neg$Bacteria was not satisfied for nearly 1% of all nuclear proteins in the PEDANT database.

The process of extracting association rules from transactions in a database consists of two phases:

1. Mining frequent itemsets from the transactions

2. Retrieving relevant rules from the frequent itemsets

Especially the phase of itemset mining is prone to combinatorial explosion, effectively prohibiting exhaustive enumeration of all possible rules. Thus, several algorithms were developed for mining association rules, which account for that issue. Some popular examples are the Apriori, Eclat and FP-growth algorithms. Two heuristics, netCAR and CPAR, are presented in detail in the Materials and Methods sections 2.5 and 2.6.

1.2.3 Support Vector Machines

SVMs are supervised machine learning algorithms [43, Chapters 6-7] [42, Chapter 5] [44, Chapter 6] [50, Chapter 9.3]. They are typically used for classification tasks, but can easily be adapted for regression as well. In support vector classification, the algorithm trains a model from given observation data, retaining a certain subset of data points (*support vectors* or *active constraints*). Using this model, new observations can subsequently be classified. SVMs as decision machines make use of non-probabilistic binary linear classifiers:

- **Linear classification**: The decision boundaries between classes are always linear (a line in two dimensions, a plane in three dimensions, a cube in four dimension, or a hyperplane in arbitrary dimensions). However, the data need not be linearly separable in input space and non-linearity can be introduced by transforming the data using basis functions (see below).

- **Binary classification**: In its basic form, an SVM searches for an optimal linear classifier that separates exactly two classes. While there are extensions that enable multiclass SVMs - creating several models in either one vs. one or one vs. all fashion - these suffer from conceptional drawbacks, including ambiguities and undefined regions.

- **Non-probabilistic classification**: Predicting new data with an SVM model yields an absolute decision of membership to one or the other class. If posterior probabilities are required, an alternative technique called *Relevance vector machine* can be used.

Support vector machines have several useful properties:

Sparse models

SVM models are sparse compared to other memory based machine learning techniques. For example, *nearest neighbor* classification uses all training points during classification. While the training phase is fast, predicting new observations is slow, because all training points need to be stored and are used during classification. In contrast, the SVM only retains a subset of training data: All training points that are closest to the decision surface are kept as support vectors (compare Fig. 1.2). Only they are required for the class prediction of new data points. All other training points can be discarded after the classifier has been found during training. This accelerates classification of new observations.

Figure 1.2: The sparse SVM model for a synthetic dataset of two classes and 30 points. The thick line indicates the classifier obtained by using Gaussian kernel functions. The margin boundary and its nine support vectors are highlighted. All other points can be discarded after training. From [43, Ch.7.1 Maximum Margin Classifiers, Fig. 7.2, p.331] with permission of Springer.

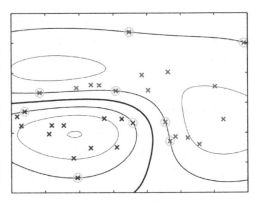

Optimal classifier

For a given classification problem, the SVM finds an *optimal* linear classifier. For an arbitrary problem of linearly separable data, a large (without consideration of machine precision: infinite) number of perfect classifiers exist (compare Fig. 1.3**A**). Many of them will, however, be overfitted to the sample they were trained on and, thus, perform badly on new observations from the population. SVMs therefore apply the concept of *maximum margins* that aims for maximal generalization of the model. The margins are defined as the perpendicular distance between the decision surface and the closest of the training points (Fig. 1.3**B**). The SVM chooses the classifier with the maximum margin. This requires an optimization procedure, which (1) identifies support vectors and (2) chooses the classifier that maximizes the margin . The problem can be formulated as a constrained convex quadratic program. Optimization algorithms can therefore find a global solution to an instance of this problem and, thus, an optimal classifier based on the maximum margin criterion.

Not linearly separable data

The considerations above assume linearly separable data and that no training points are allowed to lie within the margin. These assumptions will hardly ever hold in real applications. However, slight modifications of the SVM procedure enable its usage in scenarios of not linearly separable data. In a soft margin SVM training problem, penalized slack variables ξ_n are introduced that

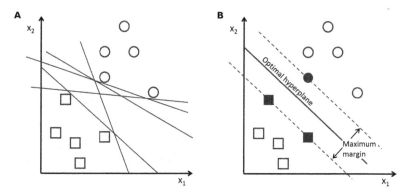

Figure 1.3: Separating hyperplanes between blue circles and red rectangles. **A**: Five linear classifiers that perfectly separate the two classes. **B**: The optimal classifier as obtained from an SVM, its margin and the three corresponding support vectors. From [51] with kind permission of Ricardo Gutierrez-Osuna.

allow *some* training points to lie within the margin or even be misclassified (compare Fig. 1.4). The parameter $C > 0$ controls the trade-off between the slack variable penalty and maximizing the margin of separation. It can be considered an (inverse) regularization coefficient that balances training error minimization and model complexity.

Nonlinear feature spaces and kernelization

In many applications, it is impossible to find good linear classifiers in input space. The canonical example is the two-dimensional Exclusive Or (XOR) classification problem (compare Fig. 1.5). Although there is no solution in input space, it is possible to linearly separate the two classes in a higher dimensional feature space. Any linear model can be extended by non-linear basis functions. In the logic XOR example, a simple quadratic feature mapping $k(\mathbf{x}_A, \mathbf{x}_B) = (1 + \mathbf{x}_A^T \mathbf{x}_B)^2$ allows for linear separation in the higher dimensional feature space, where the image of an input vector x has the form $\varphi(\mathbf{x}) = [1, x_1^2, \sqrt{2}x_1 x_2, x_2^2, \sqrt{2}x_1, \sqrt{2}x_2]^T$. The feature space is six-dimensional in this case. It can easily be shown that XOR is solvable in three dimensions. Especially for more complicated basis functions, explicitly calculating the mappings can become very expensive. Kernel functions are defined as inner products of

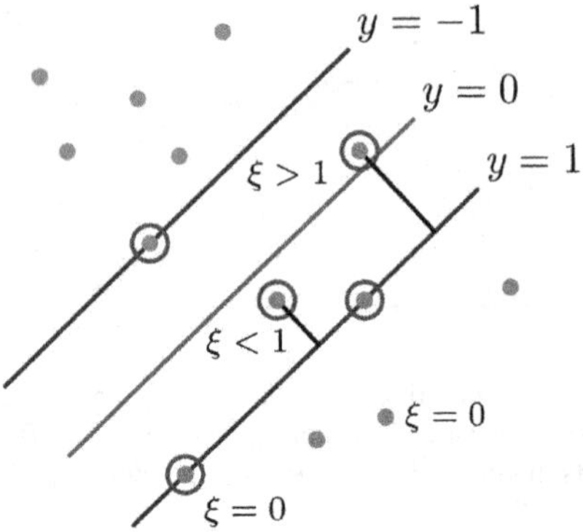

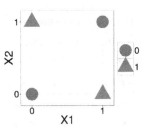

$$y = -1$$
$$y = 0$$
$$y = 1$$
$$\xi > 1$$
$$\xi < 1$$
$$\xi = 0$$
$$\xi = 0$$

Figure 1.4: Soft margin support vector classification with slack variables ξ. From [43, Ch.7.1 Maximum Margin Classifiers, Fig. 7.3, p.333] with permission of Springer.

X1	X2	t
0	0	0
0	1	1
1	0	1
1	1	0

Figure 1.5: XOR logic function: truth table and visualization. There is no linear model in two dimensions to solve this problem.

nonlinear feature space mappings $\phi(x)$. They are formally given by

$$k(\mathbf{x}_A, \mathbf{x}_B) = \phi(\mathbf{x}_A)^T \phi(\mathbf{x}_B) \tag{1.4}$$

and require the Gram matrix $\mathbf{K}$ to be positive semi-definite for all $\mathbf{x}_n$, where $\mathbf{K}$ is made from dot products of the original vectors $\mathbf{x}_n$ and $\mathbf{x}_m$, i.e., its elements are $k(\mathbf{x}_n, \mathbf{x}_m)$. Using the identity mapping $\phi(\mathbf{x}) = \mathbf{x}$ for the feature space gives the *linear* kernel

$$k(\mathbf{x}_A, \mathbf{x}_B) = \mathbf{x}_A^T \mathbf{x}_B \tag{1.5}$$

which is the simplest possible kernel function and closely related to the *polynomial* kernels, given by

$$k(\mathbf{x}_A, \mathbf{x}_B) = (1 + \mathbf{x}_A^T \mathbf{x}_B)^s \tag{1.6}$$

where s is the degree of the highest polynomial. For $s = 1$ this is the linear kernel (with a constant bias term). By strict enforcement of rule 1.4, it is possible to avoid explicit evaluations in the feature space. Instead, the complex calculations are hidden in the kernel function, which implicitly works in the feature space, without actually moving there. This is achieved by reformulating algorithms is a way, so that the feature space mappings only enter in dot product form. Then the inputs never have to be mapped to feature space, which is advantageous for high-dimensional feature spaces. For example, the most commonly used *Gaussian* kernel is defined as

$$k(\mathbf{x}_A, \mathbf{x}_B) = \exp(-\frac{\|\mathbf{x}_A - \mathbf{x}_B\|^2}{2\sigma^2}) \tag{1.7}$$

with parameter σ and is a *radial basis function (RBF)*, for which each basis function is only dependent on the Euclidean (in general: radial) distance from a center. This kernel can be shown to implicitly map the inputs to a feature space of infinite dimensionality. Without applying the concept of implicit mapping (called *kernel trick*), this would be impossible. In addition to linear, polynomial and Gaussian kernels also the sigmoidal kernel is typically available as a standard kernel in SVM toolboxes. It is defined as

$$k(\mathbf{x}_A, \mathbf{x}_B) = \tanh(\kappa \mathbf{x}_A^T \mathbf{x}_B - \delta) \tag{1.8}$$

where κ and δ are two parameters. Its Gram matrix is not guaranteed to be positive semi-definite, but the kernel is being applied, nonetheless. According to Bishop, this is "possibly because it gives kernel expansions such as support vector machines a superficial resemblance to neural network models" [43, Section 6.2, p. 299].

More complex kernels can be constructed from simple kernels following certain rules (e.g. the sum or product of two kernel functions yields another valid kernel; for more rules consult Chapter 6.2 of Bishop's book [43]). Choosing an appropriate kernel might be based on the concept of *Vapnik-Chervonenkis dimension*. More often, this is instead performed empirically by evaluating several kernels and parameters. Lin *et al.* suggest a grid-search on exponentially growing parameters and cross-validation for this purpose [52]. While SVMs profit greatly from kernelization, also other techniques can be reformulated using kernels, e.g. regression, neural networks or principal component analysis.

1.2.4 Databases for Orthologs

The webpage of the "Quest for Orthologs" consortium[2] currently lists roughly 40 different orthology databases. They are distinct in terms of taxonomic scope, number of species and methodology. There are essentially two different methods for orthology inference: Graph-based methods and tree-based methods [53]. A selection of orthology databases is given below:

- **COG** The COG database for phylogenetic classification of proteins encoded in completed genomes creates orthologous groups by merging triangles, of which each vertex is a bidirectional best hit (BBH) in pairwise protein sequences alignments. It is manually curated and based on 711 genomes as of the current release (2014 update).

- **NOG** The eggNOG database of orthologous groups and functional annotation is an extension to the COG database. Additional NOGs are created through a fully-automated

---

[2]http://questfororthologs.org/orthology_databases

pipeline. EggNOG currently features nested orthology inference across 3686 organisms. Both the COG and the eggNOG database are described in more detail in Section 2.1.

- **Ensembl Compara** is a resource for comparative genomics analyses on sequence level as well as on gene level. It describes evolutionary relationships among genes of over 300 Ensembl species.

- **OMA** The Orthologous Matrix (OMA) infers evolutionary relationships among more than 1800 complete genomes of all domains of life. It uses a graph-based method and maximum-likelihood distance estimates for orthology inference. Grouping is performed so that every pair is orthologous [53].

- **PANTHER** The PANTHER Classification System describes evolutionary relationships among genes of 104 model organisms of all domains of life. Gene functions are inferred using Gene Ontology (GO) terms.

- **TreeFam** The TreeFam database of animal gene trees performs phylogenetic reconstruction on 109 metazoans and model eukaryotes.

- **InParanoid** is a database of orthologous groups with inparalogs (homologous genes that emerge from gene duplication after a speciation event) among 273 mostly eukaryotic organisms.

- **PLAZA** is a specialized database for comparative genomics in plants and features roughly 50 Viridiplantae genomes.

- **OrthoDB** The hierarchical catalog of orthologs OrthoDB provides multiple levels of orthology. It uses a similar method for orthology inference as do the COG and eggNOG databased, which is based on merging triangles. The current version 8 features roughly 3000 species, a vast majority of which are microbes.

- **KEGG Orthology** The Kyoto Encyclopedia of Genes and Genomes (KEGG) provides manually curated ortholog groups, that allow to project experimental knowledge onto organisms, which have not been investigated so far.

The eggNOG database is used in this thesis due to the following reasons: (1) It contains a large number of microbial genomes. (2) Orthologous groups are available for several taxonomic levels. (3) EggNOG builds upon COGs and, therefore, profits partly from manual curation. (4) Recent phenotype prediction software tools utilized the database as well. Using eggNOG thus enables evaluation and comparison of these tools.

Other databases like e.g. OMA or OrthoDB are seemingly suitable for microbial phenotype prediction as well. Considering that they make use of similar methodology as eggNOG, it appears unlikely that they provide relevant advantages. Thus, only eggNOG is considered further in this thesis.

Chapter 2

Materials and Methods

2.1 Genotype data

Enumerating all genes of a species' genome is one possible level of describing its genotype. It is necessary to group genes of different species according to homology in order enable comparative genomics. Orthologous genes derive from a common ancestor through speciation events and typically share the same biological function. On the other hand, paralogs are evolutionary related through gene duplication, which enables them develop different functions. Orthologous relationships are thus of highest interest for functional inference. Genotypes are represented by clusters of orthologous groups throughout this thesis. COGs are many-to-many relationships among orthologous genes and can be considered taxonomic abstractions of genes. They are delineated from certain patterns in a graph of BBH of all pairwise protein sequence alignments: A minimal COG consists of three BBHs in a triangle. Larger COGs are built by merging those triangles that share common edges. The original COG database [54] featured 720 manually curated COGs based on seven complete genomes and was released in 1997. Due to the greater number of complete genome sequences available today, 722 genomes could be utilized for the current version (2003 COGs 2014 update), which consists of 4632 COGs.

While carefully manually curated databases like the COG database are most valuable tools for biologists, the rapidly growing amount of biological data requires (semi-)automated workflows. The eggNOG database (evolutionary genealogy of genes: Non-supervised Orthologous Groups) was created in 2007 [55] to close the gap between manually annotated and publicly available genomes. It features exact Smith-Waterman sequence alignments and automated functional annotations. The initial eggNOG release comprised 43 582 general orthologous groups extracted from 373 complete genomes across all domains of life. Roughly a quarter of these NOGs extend existing COGs. The database offers additional NOGs on hierarchically more fine-grained levels, like for example euNOGs for eukaryotes or maNOGs for mammals. EggNOG 1.0 shares its database with the STRING database of protein interactions as of version 7, which is used for the example dataset of the program NetCAR (see Section 2.5).

The updated eggNOG 2.0 was based on 630 complete genomes and yielded 69 221 coarse-grained NOGs (over 200 000 including fine-grained NOGs) [56]. EggNOG 2.0 is a sister database of STRING 8, which served as a resource for the example dataset of the program PICA [18] (see Section 2.6).

EggNOG 4.0 introduced scalable identification of high-quality genomes, quality control procedures, improved functional annotations and further enhancements [7]. The number of high-quality genomes used for NOG delineation (so-called 'core' and 'periphery' genomes) increased to 2031. Some additional 1655 genomes of poorer quality ('adherent' genomes) were also mapped to the orthologous groups. This procedure resulted in 192 421 coarse-grained NOGs. Including the NOGs from all 107 taxonomic levels, there are more than 1.7 million orthologous groups. EggNOG 4.0 data is used for evaluation purposes (Section 3.1.3) as well as for the creation of new phenotype prediction models (Section 3.2) in this thesis. COGs and NOGs will hence collectively be referred to as COGs in this thesis.

Several genomic sequences were used for this purpose that are not covered in eggNOG 4.0. These include genomes recently added to NCBI RefSeq, unpublished genomes obtained from research cooperation partners as well as metagenomic sequence bins. For these, the genotype data was extracted by the following procedure: Nucleotide sequences are processed by PRODIGAL v2.60 for gene calling using the default translation table [57]. The predicted genes are then mapped with the NCBI cognitor software [58] to an in-house generated sequence reference representing all proteins from eggNOG 4.0 COGs.

The current version eggNOG 4.1 features an improved user-interface and several technical improvements. Lacking major changes to the underlying data, it can be interpreted as an

intermediate stage before the release of the next version (presumably eggNOG 5.0), which is
expected for late 2015 and will be based on roughly 10 000 genomes.

2.2   Phenotype data

The PICA framework provides phenotype labels for its example dataset. They were obtained
from the JGI IMG [59] and NCBI Genomes [11] databases as described in [18]. Herein, the same
labels are used for reproduction and evaluation purposes of phenotype prediction tools. Ten
phenotypes are included in the example dataset. These include aerobic, anaerobic, facultative
anaerobic, gram-negative, halophilic, motile, photosynthetic, psychrophilic, endospore-forming
and thermophilic traits.

Additional phenotype information was required for novel phenotype models, which were created
for methanotrophs, nitrifiers and intracellular microorganisms. It was obtained by manual
knowledge extraction from scientific literature by the corresponding cooperation partners for
each phenotype (see Section 3.2).

2.3   Accuracy Measures

Following the definition from ISO 5725-1, *accuracy* describes a relation between *trueness* and
precision of a measurement method. Trueness is the distance from the averaged measurements
to the *true* value. Precision measures the variability within a set of measurements and thus
gives information about the reproducibility of the measurement. Measurement methods of
high trueness and high precision are called *accurate*. In binary classification tasks, accuracy
is typically defined as ratio of correctly classified cases to the total number of cases. It is
calculated as

$$\text{Acc} := \frac{TP + TN}{TP + FP + TN + FN} \tag{2.1}$$

where TP . . . true positives, FP . . . false positives, TN . . . true negatives, FN . . . false negatives. In the PICA framework, this measure is called *raw accuracy*. Under certain circumstance,

the linear raw accuracy (Fig. 2.1**AD**) is not suitable, e.g., when the classes are of heterogeneous sizes or an emphasis on either *selectivity* or *sensitivity* is required. Several other concepts exist, out of which two are presented here, as they are used by netCAR and/or PICA.

Balanced accuracy is defined as the mean of selectivity (*true negative rate*) and sensitivity (*true positive rate, recall*). It is thus a performance measure that gives penalty, if either of both terms dominates the other (Fig. 2.1**BE**).

$$\text{BalAcc} := \frac{1}{2} * (\frac{TP}{TP + FN} + \frac{TN}{TN + FP}) \qquad (2.2)$$

The *F1 score* is the harmonic mean of sensitivity and precision. It does not take TN into account, so it is dominated by the *true positive rate* (TPR). In an extreme case of only true negatives, the F1 score would evaluate to zero, even if a classifier works flawlessly (compare Fig. 2.1**CF**).

$$\text{F1} := \frac{2 * TP}{2 * TP + FP + FN} \qquad (2.3)$$

In this thesis, balanced accuracy is the measure of choice, because of ubiquitous dataset imbalance. The F1 score will be used only for the evaluation of netCAR, since it does not support balanced accuracy.

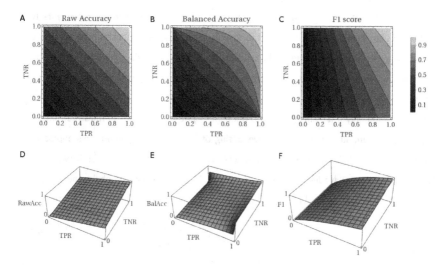

Figure 2.1: Three different accuracy measures and their linear or non-linear dependence on true positive rate (TPR) and true negative rate (TNR) depicted as contour plots and 3D plots.

2.4 KRONA

KRONA is a visualization tool that was developed for exploring taxonomy in metagenomic datasets [60]. It is nonetheless also suited for depicting taxonomy of any dataset, utilizing inherent hierarchical information of taxonomic assignments. KRONA visualizations are especially convenient when viewed on a computer, since they are highly interactive and follow *Shneiderman's mantra* (overview first, zoom and filter, details on demand). In static form they still provide a means for comprehensively depicting species of a dataset in their taxonomic context.

2.5 NetCAR

NETCAR is a heuristic CAR mining algorithm developed by Tamura and D'haeseleer [17]. It is implemented in a program named netCAR. The notation *NetCAR* will be used in this thesis for both the algorithm and the program. The algorithm was introduced for mining fixed-size sets of COGs that associate with a phenotype of interest. The program implements two additional algorithms (*Apriori, MI-CAR*) for comparative reasons. Enumeration of all possible sets quickly becomes intractable as the number of COGs and set sizes increase, as described in section 1.2.2. It is therefore necessary to sensibly reduce the hypothesis space. Standard algorithms like *Apriori* apply the *downward closure property*, which states that an itemset is frequent, if and only if its subsets are also frequent. *Apriori* thus mines all rules that have the required minimum support. In a sensitive setting of low minimum support, this approach is not efficient enough for large hypothesis spaces. NetCAR uses a COG connectivity graph to counter this problem: Nodes (COGs) are connected, if the corresponding phylogenetic profiles have sufficiently large mutual information (MI)

$$MI(X, Y) = \sum_{x,y} p_{x,y}(x, y) log \frac{p_{x,y}(x, y)}{p_x(x)p_y(y)} \qquad (2.4)$$

where $p(\cdot, \cdot)$ is a joint probability mass function and $p(\cdot)$ is a marginal probability mass function. MI is a measure of mutual dependence of two random variables. In the continuous case its unit is the *bit*, if the *logarithmus dualis* is used.

The NetCAR algorithm constructs rules by starting with single COGs (*parents*) that have MI with the phenotype above a certain threshold. Further COGs (*childs*) are considered, whose distance to their parents is less than the requested rule size. Connected subgraphs with at least one parent are subsequently considered as candidate sets. The intersection of their phylogenetic profiles is then used as the set's profile. Mutual information is finally calculated between the COG set phylogenetic profile and the phenotype profile. If it exceeds a certain threshold, a new rule is created with the COG set as antecedent and the phenotype as consequent.

NetCAR usage

NetCAR is written in the Java programming language and runs on any computer with Java Runtime Environment 1.6 or higher. A procedure with standard settings (rule antecedent size 2, NetCAR algorithm) can be invoked with the following command:

```
java -jar netCAR.jar \
./data/phylogeneticProfile.txt  ./data/phenotypeProfile.txt
```

The input file *phylogeneticProfile.txt* contains the items (phylogenetic profile) of the i-th COG in the i-th row. The j-th column corresponds to the j-th sample (species). Presence and absence are represented as 1 and 0 values, respectively. The text needs to be delimited by comma, space or tab.

phenotypeProfile.txt is formatted like the other input file. Its j-th element indicates binary phenotype classification, where 1 stands for phenotype presence, 0 for absence and -2 indicates a null value used if the phenotype status is unknown.

An output file called *2-Rule.txt* is created at the path ./data.

More information about NetCAR usage is display by the program itself, when invoked as

`java -jar netCAR.jar`

Changes to NetCAR source code

Three changes have been applied to NetCAR in order to enable the evaluation procedure:

1. Fixed re-usage of the COG connectivity graph:

   NetCAR has the option to use a pre-computed COG connectivity graph. This is especially useful, if rules for several different phenotypes are to be mined based on the same phylogenetic profiles, because the construction of this graph is a major contributor to overall computational cost. However, in NetCAR version 1.1, the main class fails to pass the connectivity graph to the NetCAR core algorithm, whose methods in turn do not take the graph as an argument. This was fixed to enable COG connectivity graph re-usage.

2. Fixed violation of transitivity principle by a comparator:

   NetCAR reproducibly throws exceptions, if confronted with rules of confidence less than 0.5. This bug is caused by the *RuleComparatorByMI* class, which implements a comparator that does not satisfy the transitivity principle $A < B \wedge B < C \Rightarrow A < C$. This was fixed by omitting one conditional.

3. Disabled undocumented output limit for 10,002 rules:

   NetCAR has no fixed limit for the number of rules it can mine in one run, however, there is an undocumented hard-coded limit for 10,002 rules that are written to the output file. While a number of rules greater than this threshold is certainly hardly ever useful, such unexpected behavior can be confusing to the user. Also, in the evaluation, I am interested in the number of rules that are created by the algorithm. This limit was consequently removed.

2.6 PICA

PICA is a software framework for phenotype prediction developed by MacDonald and Beiko [18]. It is free open source software under the Creative Commons Attribution-Share Alike 3.0 license and mainly written in the Python 2 programming language. The original release can be obtained from http://kiwi.cs.dal.ca/Software/PICA. The framework features plug-ins for different machine learning techniques. In the current release, these are association rule miners, support vector machines and a method based on mutual information. PICA provides scripts for training phenotype models from tagged data (*train.py*), predicting phenotypes of untagged data (*test.py*) and performing cross-validations for parameter selection and model quality estimation (*crossvalidate.py*). They make use of a common PICA application programming interface (API), which allows flexible usage of the framework. PICA operates on the level of presence or absence of genetic features, using COGs by default. The software has not received further improvements by the original developers since the original release in 2010. All changes I applied to PICA have been incorporated into the current release, which can be obtained from the Beiko lab website at http://kiwi.cs.dal.ca/Software/ or directly from https://github.com/univieCUBE/PICA.

CPAR

PICA implements a heuristic ARM algorithm called CPAR based on the work of Yin and Han [61]. CPAR inherits ideas from traditional rule-based classification algorithms, such as C4.5 or FOIL, as well as from ARM techniques. While the latter may yield higher accuracy, the former are often computationally cheaper. CPAR employs dynamic programming, which avoids redundancy in rule generation. FOIL uses a greedy approach to learn rules that discriminate between positive and negative examples in a dataset. It iteratively updates a rule and deletes all positive examples that are already comprehended by the current best rule. The algorithm stops, when all positives are considered. The rules are created by repeatedly adding the item p that brings the highest gain according to

$$\text{Foil gain} := |P^*|(\log \frac{|P^*|}{|P^*| + |N^*|} - \log \frac{|P|}{|P| + |N|}) \tag{2.5}$$

where $|P|$ and $|N|$ are the numbers of positive and negative examples, respectively, covered by the current rule. $|P^*|$ and $|N^*|$ indicate the values of the new rule.

CPAR modifies this greedy approach by selecting not only the best item according to Foil gain, but creates several rules at once by also considering items with slightly less gain. The newly created rules are always checked for duplicates in previously built rules. This constitutes a depth-first search in rule generation that avoids generation of a large number of candidate rules. Moreover, this procedure mines rules of variable size by design, since the algorithm only stops when a certain dataset coverage is reached or all items with sufficiently large Foil gain have already been used. This is advantageous compared to NetCAR, which requires the user to pre-define rule size.

CWMI

An additional approach to phenotype prediction in PICA is based on MI. The standard MI models similar distributions of two variables. In the biological setting genotype features and phenotypes may exhibit similar distributions for two principal reasons:

1. The feature set is a causative agent for the phenotype, thus, whenever the set is observed, the phenotype is to be expected (though not necessarily the other way round).

2. There might be a spurious relationship between the feature set and the phenotype due to common ancestry.

Taxonomy is thus a major confounding factor for phenotype prediction. Conditional mutual information (CMI) can be used to model genotype-phenotype associations under consideration

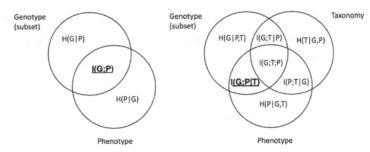

Figure 2.2: Modeling taxonomy as confounding factor in phenotype prediction with CMI. The left diagram visualizes standard MI, as the shared information between a genetic feature set and a phenotype. The right diagram introduces taxonomy as confounding factor. We search for feature sets that maximize $I(G; P|T)$, which is the amount of the MI that can not be explained by inheritance from the same ancestors. Figure taken from [18].

of taxonomic effects. It is formally defined as

$$\text{CMI}(X, Y|Z) := \sum_{x,y,z} p(x, y, z) \log \frac{p(x, y, z)p_z(z)}{p_{x,z}(x, z)p_{y,z}(y, z)} \tag{2.6}$$

where general properties match those of the classical MI (Formula 2.4, p. 30). A graphical interpretation of CMI is depicted in Fig. 2.2.

The CWMI score was developed by MacDonald and Beiko to take two aspects into account:

1. Genes of interest share high information that can not be explained by common ancestry.

2. Having identified those, we are interested in the actual MI.

It does so by scaling CMI with the difference of entropy in the phenotype variable and its MI with taxonomy and is defined as

$$CWMI(X, Y|Z) := \alpha(X, Y, Z)I(X, Y) \tag{2.7}$$

where

$$\alpha(X, Y, Z) := \frac{I(X, Y|Z)}{H(Y|Z)} \tag{2.8}$$

$$H(Y|Z) := \sum_{y,z} p_{y,z}(y, z) \log p(x, y, z) p_{y|z}(y|z). \tag{2.9}$$

Support vector classification

PICA is shipped with a plug-in for SVC. The current release features purely SVM-based prediction as well as a mixed approach utilizing ARM as a feature-selection step for SVC (called CPAR2SVM). Standard parameters are a linear kernel with soft margin cost parameter C=5. PICA makes use of the libSVM library [19]. Support vector machines are explained in detail in Section 1.2.3.

PICA usage

In this section, a typical phenotype workflow is described. It takes changes into account, which I introduced and mainly reduce the number of required command-line arguments by setting sensible standard parameters. The original PICA release necessitates more arguments. All versions give hints at program usage when called with option `python scriptname.py -h` or `python scriptname.py --help`.

Assume, we have a dataset of five species, which are represented by six genes and labeled for two different phenotypes. The input data would then comprise two files: *genotype.txt* is a file containing the genotype profile, which is essentially a binary matrix, but represented in a sparse tab-separated format: Each line corresponds to one example/species, which is identified by the first column. The following columns hold features/genes, which are present in the species. Absent genes are not stored explicitly, but will be generated by considering set of all genes in the file. Both identifier and features are represented as strings. The file content is the following:

```
Tax1 GeneA GeneB GeneC
Tax2 GeneA GeneC
Tax3 GeneB GeneD
Tax4 GeneD GeneE GeneF
Tax5 GeneA GeneC GeneE
```

The second file, *phenotype.txt*, is constructed similarly. It contains the phenotype profile, which assign YES/NO/NULL labels to each species, corresponding to phenotype presence/-

absence/unknown status. The first column contains the identifiers, which serve as primary keys and thus have to match the identifiers of the other file (arbitrary row order is allowed).

```
Species TraitX TraitY
Tax1 YES YES
Tax2 NO YES
Tax3 YES NO
Tax4 NULL NO
Tax5 NO YES
```

Having compiled all necessary input, we can now perform cross-validation in order to estimate the expected model quality. For instance, for phenotype *TraitY* this is invoked by `python crossvalidate.py -s genotype.txt -c phenotype.txt -t TraitY -o cvResult.txt > cvLog.txt`. Accuracy measures are printed to the end of *cvLog.txt*, which also contains additional information, e.g., that dimensionality could be reduced from 6 to 5, because of genes with identical phylogenetic profiles. Per-species statistics can be found in *cvResult.txt*. (Note that this is an extreme case as described in Section 2.3, where the F1-score is low despite perfect classification). The next step is training an SVM model from the whole dataset, which is done by `python train .py -s genotype.txt -c phenotype.txt -t TraitY > trainLog.txt`, which by default creates a model file named *TraitY.rules* alongside three files for conserving the class label map and the feature map (as explained in the next subsection). An optional feature ranking step processes these files and is called by `python svmFeatureRanking.py TraitY.rules > featRank.txt`, which creates a list of most predictive features:

```
Group_ID Score Class
GeneA 0.999647583602 YES
GeneC 0.999647583602 YES
GeneD -0.999647583602 NO
GeneF -0.000704832796457 NO
GeneE 6.39679281766e-18 YES
GeneB 0.0 YES
```

GeneA and GeneC are the best predictors for TraitY, while GeneD is just as informative for phenotype absence. GeneB and GeneE are not predictive for any class.

Finally, we might obtain new data like

```
Tax6 GeneD
Tax7 GeneA GeneB GeneE
```

stored in the file *novelGenomes.txt* and want to predict *TraitY*. This is achieved by invoking `python test.py -s novelGenomes.txt -m TraitY.rules -t TraitY > prediction.txt`. The following predictions are obtained:

```
Testing on 2 samples
Organism Predicted
Tax6 NO
Tax7 YES
```

Changes to PICA source code

Several changes were applied to PICA in order to enable broken functionality, add new functionality and improve general code quality. The following is a (non-exclusive) list of changes and bugfixes to PICA:

- Rewrote parts of libSVMTrainer.py and libSVMClassifier.py, so that the mapping of COGs to SVM features is correct, if COGs are present in test examples, which were not encountered in the training set. Before these changes, the SVM module could not be used for phenotype prediction of novel sequences.

  The fix makes use of *pickle* routines to save the dictionary holding the COG-feature mapping created during training alongside the SVM model to tertiary memory. In the prediction phase, this dictionary is read, so that COGs are mapped correctly.

- Hard-linked libSVM 2.90, since the library's API changed in version 3.x.

- Increased robustness of input file handling (fixed crashes on empty lines, tabs at *End Of File*, etc. in genotype and phenotype files).

- Improved error handling (message to the user, if input files do not exist or some arguments were not passed to the program, instead of crashing).

- Changed test.py, so that it does not require a phenotype input file. Unless used in a validation scheme, the phenotype profiles of the genomes to be classified are unknown. The requirement is therefore meaningless in actual phenotype prediction scenarios.

SVM feature ranking

In addition, I extended the software by implementing a function to extract the most predictive features from linear SVM models based on weights as described by Chang and Lin [62]. The source code of the Python script `python svmFeatureRanking.py`, can be found in the supplement (Listing A.1, p. 97). Dumping and ranking features enables the user to interpret the models from a biological perspective: the COGs with the highest discriminative power for presence or absence of a phenotype are ranked at the top. If specific proteins are already known to be part of the genetic underpinning of a phenotype of interest, the user finding the according COGs ranked highly in the list might gain confidence in the biological relevance of the predictions. Unexpected high-ranking COGs hint at previously unknown protein functions associated with the phenotype. However, these may also derive from taxonomic correlations rather than a shared phenotype, which can be caused by insufficient training data.

COGs with identical scores were treated as one feature by PICA, because they have identical phylogenetic profiles. This must be taken into account, when an SVM model shall be interpreted with respect to causal relationships between features and phenotypes.

2.7   GenTraitor

GenTraitor (*Generic Trait Predictor*) is the working title for an integrated phenotype prediction pipeline. The goal is to develop software that automatically performs all necessary steps for phenotype prediction and only requires DNA sequence information (genome, metagenomic bin) as input. The first step of this process is annotation of genotypic features. As of now, this involves gene calling with PRODIGAL v2.60 using the default translation table [57] and mapping them with the NCBI cognitor software [58] to an in-house generated sequence reference representing all proteins from eggNOG 4.0 COGs. This procedure requires expensive PSI-BLAST [63] computations. Other procedures might be implemented to decrease computational cost. For example, Hidden Markov model (HMM) profiles might be used to search for COGs in the sequence data directly. Alternative concepts of genotypic features, like for instance Pfam protein families [64] could be utilized as well and are currently investigated by a colleague. Next, the pipeline would check the genotype data for genome completeness, e.g. based on marker genes or more sophisticated tools like CheckM [65] for metagenomic data. This information may be used to exclude those models that are known to perform bad for the given completeness value or to give warnings about such issues. The following step is the actual phenotype prediction, which is performed by PICA for all selected models. Finally, the predictions are presented to the user. While this pipeline can be used on demand, it would also be possible to broaden the scope and classify for example all genomes deposited in the RefSeq database. Ideally, the results of these predictions would finally be deposited in a publicly available online database. At the moment, the different tasks have not yet been integrated as GenTraitor, but are still executed manually.

Chapter 3

Results and Discussion

3.1 Comparison and Evaluation of Phenotype Prediction Packages

3.1.1 NetCAR

Initially, I set up the phenotype prediction software netCAR. This package provides heuristic association rule mining, based on similarities in phylogenetic profiles of COGs. I fixed several issues as described in section 2.5. After applying these changes, I was able to run the software and to reproduce results from the original publication. NetCAR was executed as to extract predictive COG-sets of size 1 to 5 for six different phenotypes: aerobic/anaerobic/facultative anaerobic lifestyle, endospores, gram-negativity and motility. The input data comprised 155 prokaryotic genomes, represented by their COG profiles, which were based on the eggNOG database version 2. Overall, 11969 different COGs were present in those profiles. Exhaustive enumeration of all distinct COG subsets is impossible due to combinatoric explosion. While this dataset is reasonably small, the number of unique COG sets is tremendous as can be seen in Table 3.1. Therefore, heuristics such as netCAR have to be applied, in order to reduce the hypothesis space.

k	Number
1	11 969
2	71 622 496
3	2.8570×10^{11}
4	8.5468×10^{14}
5	2.0452×10^{18}

Table 3.1: Number of distinct k-combinations of 11 969 COGs

Mining rules of variable size

However, this computational issue is worsened again by the fact that netCAR only mines rules of a fixed size in each run. If, for example, rules of size 2, 3 and 4 are to be calculated, netCAR must be called three times. These runs are independent, so the program can not profit from having calculated smaller rules before, that are already informative for larger rules. The size of a good predictor is generally not known a priori. For that reason, running netCAR only once is often insufficient for finding optimal rules. For most phenotypes, the number of predicted rules grows exponentially for increasing rules sizes (Table 3.2). The number of unique COGs within these rules grows, however, approximately linearly. Thus, there is a high level of redundancy between the rules and much computation time is wasted on evaluating highly similar rules over and over again.

	aerobic		anaerobic		endospore		gramnegativity		motility	
k	Rules	#	Rules	#	Rules	#	Rules	#	Rules	#
1	2	2	16	16	61	61	89	89	24	24
2	13	11	11	12	72	55	3341	324	538	79
3	37	11	11	12	440	69	147379	505	13377	261
4	260	24	10	6	7291	96	34017039	624	3700746	408
5	3686	39	6	6	6493956	321	NA	NA	NA	NA

Table 3.2: Number of predictive rules (k-sets) and number of unique COGs (#) within these rules per phenotype. NetCAR did not predict any rules for the facultative anaerobic phenotype. NA values indicate that the corresponding computation was aborted after 24 hours.

Association rules for six exemplary phenotypes

Exemplary results of the netCAR evaluation are shown in Table 3.3. The algorithm performs well on prediction of the gram-negative trait and reasonably well for endospores and motility.

However, the dependency on an oxygenic environment is harder to grasp. NetCAR was not able to predict any rules for the facultative anaerobic phenotype. The differences in prediction accuracy seem to be well reflected in the COG descriptions. The presence of an adenylate kinase is predictive for the motility trait. ATP-driven motility based on flagella has high energy consumption. Dependencies on adenylate kinases are known for both prokaryotic and eukaryotic motility [66][67]. Thus, a relation between motility and such kinases is plausible (although on its own unspecific, since other cellular mechanisms depend on adenylate kinases as well). The less accurate aerobic rule contains a ribosomal protein. Its corresponding COG is one of 40 universal markers for prokaryotes as described in [68]. As such, the protein can not be predictive for any phenotype. The general performance of netCAR is assessed in parallel to its direct competitor PICA in Chapter 3.1.2.

Phenotype/ F1 score	ID	COG description
aerobic	COG0081	Ribosomal protein L1
73.7	COG0437	Fe-S-cluster-containing hydrogenase components 1
	COG0751	Glycyl-tRNA synthetase, beta subunit
anaerobic	COG0366	Glycosidases
63.2	COG1041	Predicted DNA modification methylase
	COG1621	Beta-fructosidases (levanase/invertase)
facultative anaerobic	NA	
endospore	COG1194	A/G-specific DNA glycosylase
81.8	COG1359	Uncharacterized conserved protein
	COG2192	Predicted carbamoyl transferase, NodU family
gramnegativity	COG0136	Aspartate-semialdehyde dehydrogenase
95.7	COG1077	Actin-like ATPase involved in cell morphogenesis
	COG4345	Uncharacterized protein conserved in archaea
motility	COG0563	Adenylate kinase and related kinases
84.6	COG0695	Glutaredoxin and related proteins
	COG0748	Putative heme iron utilization protein

Table 3.3: Highest ranked rules of size 3 for each phenotype and their F1 score. The description for each COG is given. NetCAR did not predict any rules for the facultative anaerobic trait.

3.1.2 PICA

The PICA framework comes with example data that was used in the original publication. I used these data to conduct a two-fold cross-validation on ten different phenotypes to check for proper functioning of the software (Fig. 3.1).

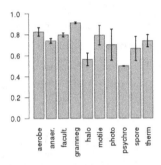

Figure 3.1: CPAR 2-fold cross-validation for ten different phenotypes:
aerobe...aerobic,   anaer...anaerobic,   facult...facultative   anaerobic,   gramneg...gram-negative,   halo...halophilic,   motile,   photo...photosynthetic,   psychro...psychrophilic, spore...endospore-forming, therm...thermophilic.
Error bars indicate standard deviation of F1 scores.

CPAR outperforms netCAR

The PICA framework [18] utilizes CPAR as an alternative ARM heuristic. A cross-validation scheme was used to compare its performance with netCAR. Ten replicates of training and test sets for a 5-fold cross-validation were derived from PICA example data. For each training set, association rules for ten phenotypes were mined by both netCAR and PICA and converted to PICA-format, if necessary. PICA's test core was subsequently used to predict these phenotypes in the test sets. The performance was measured in terms of the F1-score (Fig. 3.2) as well as the required runtime for training (Fig. 3.3). The CPAR algorithm outperforms netCAR in terms of prediction quality in seven out of ten cases. Its median F1-score over all phenotypes is 0.04 greater than netCAR's median F1-score. While being slightly more accurate, CPAR is also considerably faster. The median training time is approximately 120 times longer using netCAR. Thus, I refrain from employing netCAR further.

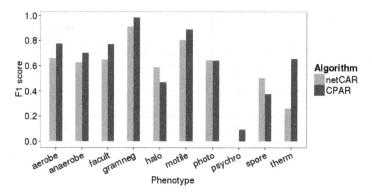

Figure 3.2: Comparison of association rule heuristics: prediction quality in 10x 5-fold cross-validation.

Classification with support vector machines

In addition to ARM the PICA framework also provides a binding to libSVM [19], a widely used toolbox for support vector machines. There are two different usage scenarios in PICA that utilize SVMs:

1. Prediction of association rules with the CPAR algorithm as a feature-selection phase for subsequent SVM learning (referred to as CPAR2SVM from now on). Only those features are used for support vector classification, that are elements of at least one of the rules mined by CPAR. Thus, the feature space is the union of all association rule antescedents.

2. Direct application of support vector classification without feature-selection phase (referred to as SVM). All features (orthologous groups) present in the input data are used in the SVM.

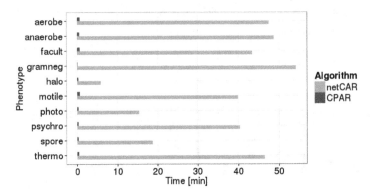

Figure 3.3: Comparison of association rule heuristics: runtime per fold and replicate (training only).

I applied a 5-fold cross-validation scheme with ten replicates to compare the prediction quality of the three different classification algorithms provided by PICA (Fig. 3.4). Both support vector classification based approaches outperform ARM in all cases. There is no significant difference between those two in terms of the obtained balanced accuracy. However, I observed approximately three-fold longer runtimes for CPAR2SVM than for SVM (median over ten phenotypes, Fig. 3.5). Hence, I choose PICA with SVM as the standard procedure for further phenotype prediction.

Surveying different SVM kernels

Support vector classification was done with PICA standard settings (linear kernel, soft margin cost parameter $C = 5$). While being computationally less expensive, the linear kernel is supposed to be inferior to a Gaussian RBF kernel in general [69]. To assess the influence of different kernels on prediction quality, I performed cross-validation (same as before) with four different common kernels, using standard parameters, where they apply (Fig. 3.6). In all cases, the linear kernel performs best. It is expected that extensive grid search for the best parameters for the RBF kernel should improve its performance to match the linear kernel. This procedure would have to be performed for each new phenotype model. It is potentially expensive, since it requires cross-validation for each parameter tuple. Therefore, I retain the linear kernel as

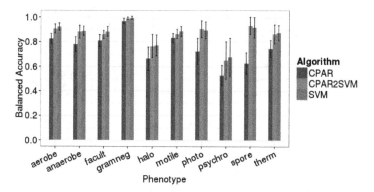

Figure 3.4: Comparison of ARM- and SVM-based methods: prediction quality in 10x 5-fold cross-validation. Error bars indicate standard deviation.

standard setting in support vector classification while maintaining the possibility to use other kernels for complex traits, which might fail to be grasped by a linear model.

3.1.3 Complexity Estimation

Up to this point I evaluated PICA based on the example data set the framework is supplied with. It comprises COG profiles derived from eggNOG 2.0 [56]. The eggNOG database received several updates since the release of PICA. The current version is 4.1 as of July 2015. This section is based on version 4.0 [7], because the experiments were performed ahead of the eggNOG 4.1 release in May 2015. The new versions brought several changes to the database. One important difference for my purposes is the increasing number of orthologous groups overall as well as specifically NOGs on LUCA level. Considering all taxonomic levels, the number of orthologous groups increased from approximately 225 000 to more than 1 700 000. However, all experiments were performed solely on COGs of the highest taxonomic level (unless stated otherwise explicitly), which leads to a problem dimensionality much smaller than it may look at a first glance. The number of LUCA-based COGs present in at least one of the species from the example data set increased from 47 615 to 192 421. Phenotype prediction was thus repeated for the ten example traits in this updated genotype space, while re-using the

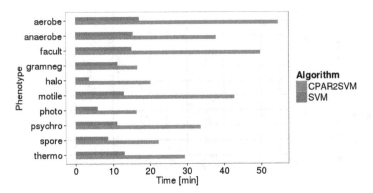

Figure 3.5: Comparison of SVM-based methods: runtime for a complete 10x 5-fold cross-validation (training and testing).

original phenotypic trait profiles. With this scheme we can estimate the impact of increasing dimensionality on the required computational resources. The average computation time for the standard cross-validation procedure increased from 12 minutes to 43 minutes (Fig. 3.7). Higher computational costs are not surprising in this scenario and, indeed, the rise is only moderate. Due to this we consider phenotype predictions based on eggNOG 4.0 profiles computationally tractable. For some phenotypes significant differences were observed in terms of prediction accuracy (Fig. 3.8). There is a trend of reduced accuracy in the vaster genotype feature space of the newer eggNOG version. The median over the average prediction accuracy decreased from 0.887 in version 2.0 to 0.862 in version 4.0.

Assessing bactNOGs as genotype space

Orthologous groups from the eggNOG database are not restricted to LUCA level. Besides this, there are 106 more fine-grained levels available, ranging from e.g., arNOGs (domain of *Archaea*) to levels as deep as arthNOG (familiy of *Arthrodermataceae*). I evaluated phenotype prediction in the bactNOG space which is made of orthologous groups within the domain of *Bacteria*. Species and phenotype profiles for the ten example traits were retained from the previous experiments. Predictions with this approach are as accurate as predictions in LUCA level COG space or slightly less accurate (Fig. 3.9). Utilizing bactNOGs restricts to do

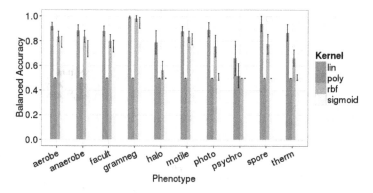

Figure 3.6: Prediction quality for different SVM kernels: lin ... linear, poly ... polynomial, rbf ... radial basis function, sigmoid. Error bars indicate standard deviation.

predictions within one domain of life only. Because of the limited scope and the non-increasing accuracy, I decide not to employ bactNOGs any further, but use the general NOGs on LUCA level for all other prediction scenarios.

Increasing sizes of training data sets

Besides the ability to cope with a high number of features (COGs), it is also crucial that the software can deal with many data points (genomes). While increasing dimensionality for evaluation of phenotype prediction packages is straight-forward and arises naturally from eggNOG usage as seen before, providing more data is generally hard. Each data point represents a genome and, consequently, a species. While more and more genomes are made available, there is a general lack of information about correspondence to phenotypes. It is therefore impossible to increase the number of genomes arbitrarily, because of insufficient biological knowledge. Therefore, I use a different approach to assess the effect of growing numbers of available genomes on computation time and memory consumption.

Virtual species were created as surrogates for novel biological data: I defined a set of possible features (virtual COGs) and selected subsets (virtual genomes). Their sizes are variable and correspond to typical microbial genome sizes. Virtual trait labels were assigned to the genomes randomly. I compiled several training sets of increasing numbers of genomes and performed

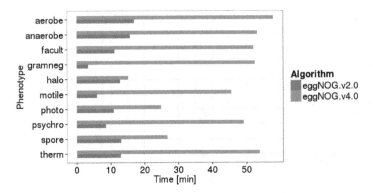

Figure 3.7: Runtime for 10x 5-fold cross-validations using different versions of eggNOG.

cross-validations on them.

With growing problem sizes, both runtime and memory usage increase. However, the results indicate that 5000 genomes comprising a total of approximately 200 000 different COGs in the training set are computationally feasible (Fig. 3.10). I consider this a "worst case" scenario. Typical phenotype predictions involve not more than several hundred genomes with labels, depending on current biological knowledge about the traits of interest. Furthermore, I expect the dimensionality not to rise dramatically. The number of COGs is theoretically bounded by the genome size of LUCA.

Hence, I deem PICA capable of handling large data sets that will arise in the near future, especially from metagenomic studies.

3.1.4 Prediction on Incomplete Genomes

Phenotype prediction needs to be robust against genome incompleteness, if it is to be applied to metagenomic data. Binning of metagenomic sequences usually results in incomplete genomes due to limitations and prediction errors of assembly and binning algorithms themselves. Thus, prediction accuracy in metagenomic contexts is a function of genome completeness.

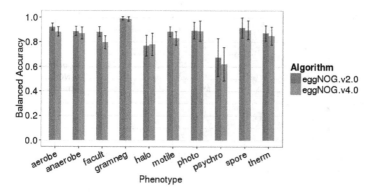

Figure 3.8: Prediction accuracy in 10x 5-fold cross-validations using different versions of eggNOG.

A strategy for assessing the incomplete genome case

I investigated this property with simulated incomplete genomes. Initially, a performance baseline was established for each of ten example traits by conducting cross-validations with all available genomes. All of them are complete genomes. Genotype data was obtained from eggNOG 4.0. The results are estimations of the highest accuracy that can be achieved subject to the given data per trait. The performance on metagenomic bins was then assessed by simulating incomplete genomes. Additional cross-validations were performed in which models were built from training sets comprising complete genomes. For the test sets, each genome was simulated for incompleteness by randomly deleting a fixed fraction of COGs. As an illustrative example, to simulate 70 percent completeness from a genome of 1000 COGs , 300 COGs are selected at random and subsequently removed from the data set. Besides the complete genomes scenario, nine levels of completeness were simulated from 10 to 90 percent. Three replicates were created for the random removal of COGs per genome and completeness level. Consequently, phenotype prediction was performed on simulated incomplete genomes with models derived from complete genomes.

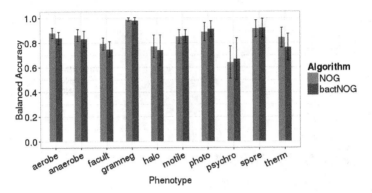

Figure 3.9: Prediction accuracy in 10x 5-fold cross-validations using orthologous groups on different taxonomic levels.

Phenotype prediction is feasible for incomplete genomes

For ten example traits sigmoidal response curves are observed for increasing genome completeness (Fig. 3.11). Prediction accuracy saturates at 60-70 percent completeness for most traits. The models for gram-negativity perform best overall and are still useful in completeness regimes around 30-40 percent. The below-average models for psychrophily are primarily restrained by the small number of species in the training set known to give rise to this trait (17 positives). As for the other traits, only endospore-formation and photosynthesis models profit notably from genome completeness above 70 percent. Balanced accuracy below 50 percent is not observed, because the SVM-based methodology is prone to assign all genomes to one class when confronted with insufficient data. This class is usually phenotype absence, if the trait is primarily predicted through feature presence. Thus, specificity of 1.0 and sensitivity of 0.0 (or vice versa) arise, which necessarily result in a balanced accuracy of 0.5.

Given the results of this experiment, I expect that meaningful phenotype predictions can be performed on metagenomic bins, even if they represent incomplete genomes.

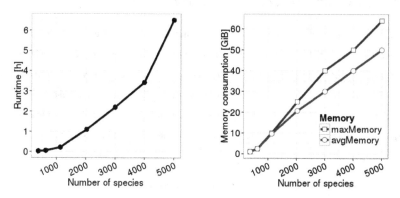

Figure 3.10: Runtime and memory consumption for a single fold of cross-validation for increasing problem sizes. Problem dimensionality increases with the number of species to approximately 200 000 for 5000 species. Memory was measured as peak main memory (maxMemory; needed only for a short period after start of a cross-validation) and average main memory (avgMemory; constant requirement after the peak).

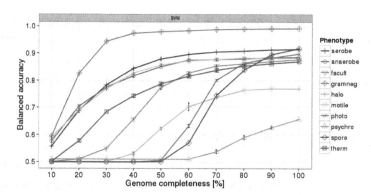

Figure 3.11: Phenotype prediction performance for incomplete genomes. Each point represents a cross-validation with training on complete genomes and testing on incomplete genomes. Incompleteness was simulated by random removal of x percent of all COGs in a genome. Error bars indicate standard deviation of three replicates of random COG removal.

3.2   Extension to Novel Phenotypes

Given the successful evaluation of PICA-based phenotype prediction, I extended the study to additional traits. Initially, methane oxidation and nitrification were investigated. These are simple metabolic traits, for which single characteristic marker genes are known. I also tested the prediction framework for genetically more diverse traits, like for example intracellular lifestyle. Further models for bacterial secretion systems are currently being created by a colleague in the group. The training data for models in this chapter were compiled by experts of each area, namely

- Methanotrophs: Alexander T. Tveit (Arctic University of Norway)

- Nitrification: Holger Daims (Division of Microbial Ecology (DoME) at the University of Vienna)

- Intracellular lifestyle: Frederik Schulz and Matthias Horn (also DoME)

3.2.1   Methane Oxidation

Biological methane oxidation is performed by specialized prokaryotes, called methanotrophs, that can utilize methane as energy and carbon source. Methane monooxygenases (MMOs) are key enzymes in this process, that catalyze methanol production from methane [70]. The particulate methane monooxygenase gene *pmoA* has long been used as a phylogenetic marker for methanotrophs [71].

Creating a model for methanotroph prediction

I first trained a model with 33 known methantrophs (see Fig. 3.12) and 6 closely related non-methanotrophs. Using these species as input data, a standard cross-validation yielded a balanced accuracy as low as 0.627 ± 0.210. This result is caused by a high number of false positives, which is most likely due to the very low number of negative training data. In order to counter this problem, my cooperation partner provided 80 additional non-methanotrophs (see Fig. 3.13), which I used to create a new model. Cross-validation for the increased training set

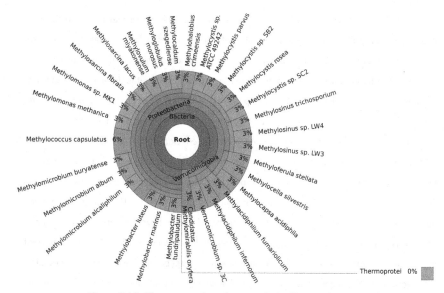

Figure 3.12: Taxonomy of methanotrophs in the training set.

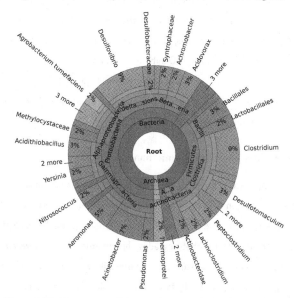

Figure 3.13: Taxonomy of non-methanotrophs in the training set.

resulted in balanced accuracy of 0.972 ± 0.046. A total of 112 species was predicted correctly in all replicates, while only 7 species were at least once misclassified. One single false negative was observed: *Candidatus Methylomirabilis oxyfera* was not recognized as methanotroph in any replicate. We hypothesize, that this may be caused by unusual metabolic features, especially the absence of a complete denitrification pathway [72]. Two *Methylacidiphilum* species and *Verrucomicrobium sp. 3C* were misclassified in only one replicate (of a total of ten). In the group of closely-related non-methanotrophs *Beijerinckia indica ATCC9039* was always incorrectly predicted as methanotroph. This misclassification might be due to a putative propane monooxygenase homologous to soluble propane/methane monooxygenases [73]. Two *Nitrosococcus* species were misclassified in four replicates. All of the non-closely related non-methanotrophs were always predicted correctly. I conclude that adding distantly related species as negatives to the training set can tremendously increase the predictive power of models in cases of few available close relatives. One can speculate that distant relatives allow the algorithm to find a genomic "baseline", which consists of genes independent of the specific trait. On the other hand, closely related negatives might then allow for a more pronounced discrimination based on genes that are not part of the baseline. The most predictive features of the methanotroph model were calculated with the feature ranking algorithm (see Table 3.4). Three of the top ten features code for MMO subunits with *pmoB* on rank 1 and *pmoA* on rank 4. *Mxa* genes code for subunits of methanol dehydrogenases (MDH), which are ubiquitous in gram-negative methylotrophs [74]. Several MDH chains are also ranked prominently among the 30 most predictive methanotroph features.

Searching for previously unrecognized methanotrophs

The methanotroph model was subsequently used to predict this trait in all species of eggNOG 4.0 (core and periphery genomes). Eight bacteria are predicted as methanotrophs, seven of which are also present in the positive training set. One additional prediction is *Hyphomicrobium sp. MC1*, an aerobic methylotroph, that degrades chloromethane [75]. However, it lacks the characteristic methane monooxygenase and is, thus, most likely incapable of oxidizing methane. While the model is generally highly specific for methanotrophs, their discrimination from related species performing similar metabolic functions is still suboptimal.

Feature ranking for methanotroph model

Rank	COGID	Score	EggNOG description
1	NOG11046	0.01210	monooxygenase, subunit B
2	NOG00933	0.01201	selenium-binding protein
3	NOG01206	0.01183	tonB-dependent Receptor
4	NOG24232	0.01128	methane monooxygenase ammonia monooxygenase subunit A
5	NOG66789	0.01072	
6	COG1795	0.01051	Formaldehyde-activating enzyme
7	COG3252	0.01051	Catalyzes the hydrolysis of methenyl-H(4)MPT( ) to 5- formyl-H(4)MPT (By similarity)
8	NOG58483	0.01051	MxaA protein
9	NOG08478	0.01008	Methane monooxygenase ammonia monooxygenase, subunit C
10	NOG11089	0.00995	von Willebrand factor type A
11	COG2218	0.00954	formylmethanofuran dehydrogenase, subunit C
12	COG0444	-0.00950	(ABC) transporter
13	NOG03854	0.00938	Dehydrogenase
14	NOG40434	0.00938	Protein of unknown function (DUF447)
15	COG2037	0.00923	Catalyzes the transfer of a formyl group from 5-formyl tetrahydromethanopterin (5-formyl-H(4)-MPT) to methanofuran (MFR) so as to produce formylmethanofuran (formyl-MFR) and tetrahydro-methanopterin (H(4)MPT)
16	COG4993	0.00921	Dehydrogenase
17	COG2010	0.00874	cytochrome c
18	NOG13123	0.00866	(ABC) transporter
19	COG3383	0.00861	Formate dehydrogenase Alpha subunit
20	COG0627	-0.00848	esterase
21	NOG42857	0.00844	MxaD protein
22	NOG53644	0.00844	MxaD protein
23	NOG64838	0.00844	Methanol dehydrogenase beta subunit
24	COG1029	0.00841	formylmethanofuran dehydrogenase subunit B
25	COG1229	0.00841	formylmethanofuran dehydrogenase, subunit a
26	COG1548	0.00841	H4MPT-linked C1 transfer pathway protein
27	NOG43152	0.00835	cytochrome C, class I
28	NOG79346	0.00821	MxaK protein
29	NOG05324	0.00820	von Willebrand factor type A
30	COG2259	0.00812	doxx family

Table 3.4: Top 30 most predictive features for the methanotroph trait as obtained from the feature ranking algorithm. Positive and negative scores indicate features that are predictive for phenotype presence and absence, respectively.

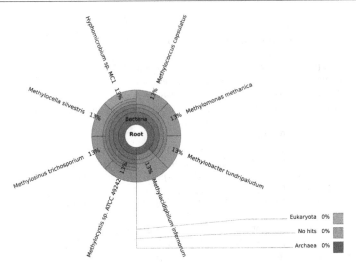

Figure 3.14: Prediction of methanotrophs in eggNOG 4.0

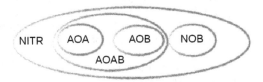

Figure 3.15: Hierarchy of nitrification models. Abbreviations as defined in the main text. AOAB . . . ammonium oxidizing prokaryotes, NITR . . . nitrifiers.

3.2.2 Nitrification

Nitrification is an important process in the biochemical nitrogen cycle in soil [76]. The oxidation of ammonium to nitrite is carried out by bacteria (AOB) and archaea (AOA). Subsequent oxidation of nitrite to nitrate is performed by specialized bacteria (NOB). Key enzymes in this process are ammonium monooxygenase (amo) and nitrite oxidoreductase (nxr). Their subunits *amoA* and *nxrB* serve as markers for nitrifying microbes [16][77].

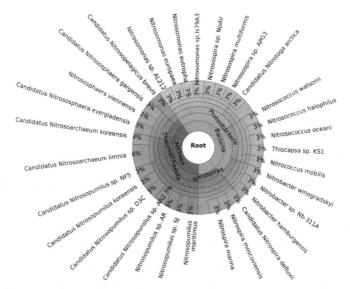

Figure 3.16: Taxonomy of nitrifiers in the training set.

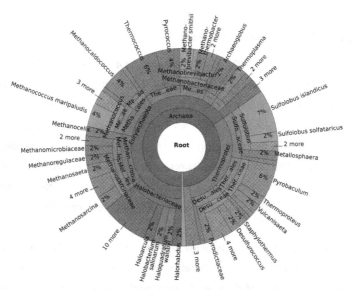

Figure 3.17: Taxonomy of non-nitrifiers in the training set.

Trait	Accuracy	SD
AOA	1.000	0.000
AOB	0.996	0.030
AOAB	0.992	0.041
NOB	0.873	0.159
NITR	0.977	0.047

Table 3.5: Prediction accuracy of nitrification models in cross-validation (mean balanced accuracy, standard deviation).

Five shades of nitrification

In the context of phenotype prediction, one important issue is the definition and delimitation of a single phenotypic characteristic. For instance, nitrification as a whole can be considered one characteristic. However, it can also be conceived as an abstraction of several characteristics, namely nitrite oxidation and ammonium oxidation, which in turn may be broken down into AOA and AOB. Therefore, I created several models for nitrification on all levels, i.e. nitrification, nitrite oxidation, ammonium oxidation, AOA and AOB. The hierarchy is depicted in Figure 3.15.

Creating models for nitrification prediction

The training set consists of 11 NOBs, 14 AOAs and 10 AOBs, adding up to 24 ammonium oxidizers and 35 nitrifiers altogether. Their taxonomy is depicted in Figure 3.16. The training set also comprises 337 non-nitrifiers (Figure 3.17). Cross-validation with these five training sets results in high accuracy (Table 3.5). The prediction of AOA is performed flawlessly in all folds. However, all the AOAs in the training data are Thaumarchaeota, and all known species in this phylum are AOA. Without closely related negatives, models might rather reflect common ancestry than a shared trait. Indeed, no nitrification marker genes can be found among the most predictive feature. I thus conclude, that the AOA model predicts Thaumarchaeota in general.

Also the prediction of AOB results in nearly maximal accuracy. Only three *Nitrosomonas* species are misclassified once in ten replicates of cross-validation. In this case, both AOB-positive and negative the class of *Betaproteobacteria* and the order of *Chromatiales* are present in the training set. This seems to enable the support vector machine to discern differences between phenotypic signals and purely taxonomic ones. The most predictive feature in the AOB model is in fact *amoB*, coding for the beta chain of ammonium monooxygenase. The marker gene *amoA* is ranked 89th.

Prediction of nitrite oxidation is less accurate than prediction of ammonium oxidation. This is mostly due to three species being misclassified in all replicates: *Nitrococcus mobilis Nb-231*, *Nitrotoga arctica 6680* and *Thiocapsa KS1*. The latter is of high interest, because it is a phototrophic NOB [78], deriving energy from light instead of oxidation of environmental electron donors. This might be an important factor for the misclassification. The marker gene *nxrB* is ranked first among the most predictive features.

A highly predictive uncharacterized protein

The model for the combined nitrification trait performs highly in cross-validation, with a mean balanced accuracy of nearly 98 %. In this case, the most predictive feature is NOG83329, representing a domain of unknown function (DUF2024). This NOG also ranks highly in the individual AOA, AOB and AOAB models, but seems to play no major role for NOB. One can speculate, that this domain might be related to ammonium oxidation. Not much is known about DUF2024, besides an NMR-structure (PDB accession 2HFQ). Analysis with the TopSearch web service [79], however, reveals structural similarities to nitrogen regulatory protein P-II and nitrogen fixation protein NifU. The former is a known activator of glutamine synthetase in *Mycobacterium tuberculosis*, which converts glutamate and nitrate to glutamine. While correlation of a possibly similar function with the nitrification trait appears to be plausible, the actual role of DUF2024 remains unclear.

Searching for previously unrecognized nitrifiers

I used all five models in order to predict nitrification for all genomes in eggNOG 4.0 (core and periphery). A total of 16 species are predicted general nitrifiers (see Fig. 3.18), out of which there are five predicted NOB and eleven ammonium oxidizers (three AOA, eight AOB). On each trait level the number of predictions sum up to the number of the next higher level and

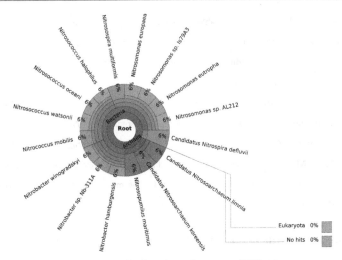

Figure 3.18: Predicted nitrifiers in eggNOG 4.0

there are no overlaps on same levels. The results of the three trait levels are thus compatible with respect to each other. All of the predicted nitrificants are also present as positives in the training set. There are no unexpected predictions. Therefore, all five models predict the corresponding nitrification trait with high specificity.

3.2.3   Intracellular lifestyle

So far I have investigated the phenotype prediction of traits, which can be inferred mainly from the presence of certain proteins. For instance, genes encoding particular porins are essential for the outer cell membrane of Gram-negative bacteria, or genes encoding specific regulators are characteristic for genomes of spore-forming microbes. However, phenotypes might also be determined by the absence of genes.

Gene absence can be used for phenotype inference

In order to assess the possibility of predicting such traits, I investigated obligate intracellular lifestyle in bacteria. The working hypothesis is that this trait should largely be predicted by

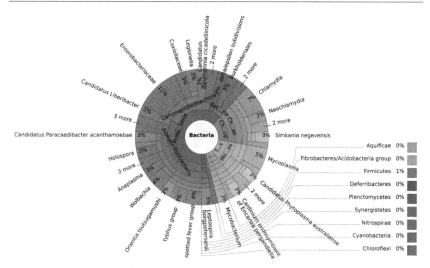

Figure 3.19: Taxonomy of intracellular bacteria in the training set.

absence of genes in the reduced genomes of such microbes. I trained a model for obligate intracellular against free-living species and facultative intracellular species. The training data comprise 43 obligate intracellular, 6 facultative intracellular and 48 free-living bacterial species. Their taxonomic diversity is depicted in Figs. 3.19 and 3.20. The data amounts to a total of 97 genomes, containing 30455 unique COGs. Only complete genomes were considered in this case to enable inference based on feature absence. Genomes were deemed complete if at least 39 out of 40 universal prokaryotic marker COGs [68] were present. The assignment of completely sequenced genomes to obligate intracellular, facultative intracellular, and free-living phenotypes was performed by manual knowledge extraction from scientific literature (such as [80] [81] [82] [83] [84]) by the cooperation partners. Cross-validation was performed to estimate the model quality. I observed a prediction accuracy of 0.997 ± 0.010, which is all positives as well as all negatives were predicted correctly in all cases, except for *Cand. Pelagibacter sp. IMCC9063*. This free-living organism with a small streamlined genome, which indeed shows some features of obligate intracellular microbes [85], was misclassified as obligate intracellular in 3 out of 10 folds.

Subsequently, I created a model from all 97 species with intracellular labels and performed

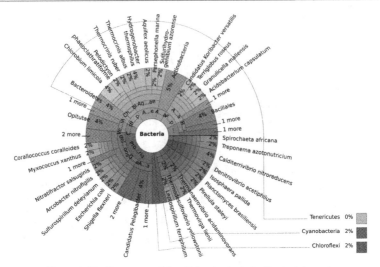

Figure 3.20: Taxonomy of free living bacteria in the training set.

feature ranking. All top 50 features are negative predictors, i.e., their absence is predictive for the obligate intracellular trait (Table B.2, p. 108). Only 2 positive predictors are present in the top 100 and 14 in the top 200 (data not shown). Hence, predicting this phenotype is indeed primarily based on gene absence.

Searching for previously unrecognized intracellular microbes

I tested all bacterial species in eggNOG 4.0 (core and periphery genomes filtered for marker COGs as described above) for obligate intracellular lifestyle. A total of 160 species is predicted intracellular (Figure 3.21). As expected, many known intracellular bacteria were found in our analysis, e.g., several *Mycoplasma*, *Chlamydia*, *Borrelia* and *Rickettsia* species or *Coxiella burnetti*. Also six *Bartonella* species were predicted obligate intracellular although they are truly facultative intracellular and can be cultivated in vitro. Yet they are highly fastidious and show genome features indicative of a host-integrated metabolism [86]. It is evident, that the discriminative power of the model between obligate and facultative intracellular species is suboptimal so far. This is not surprising as there is a smooth transition between these life styles, largely blurred by our ability to meet growth requirements or simulate intracellular con-

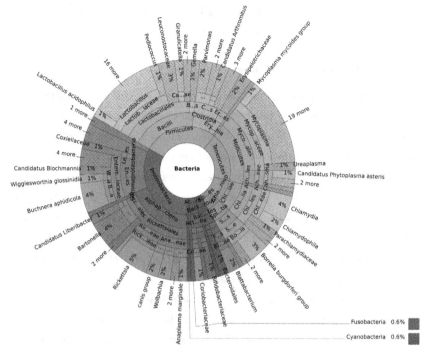

Figure 3.21: Taxonomy of predicted obligate intracellular bacteria in eggNOG 4.0. All species were considered, whose genomes are flagged as complete by considering 40 universal marker COGs.

ditions in the lab for some microbes that are host-associated in nature. In addition, the very small number of facultative intracellular organisms (six bacteria) in the training set further aggravated this distinction. A number of free-living Firmicutes, including several *Lactobacillus* species, are examples for predicted intracellular microbes, for which the reason for their unexpected classification still needs to be deciphered. Several known intracellular bacteria present in eggNOG 4.0, like for example *Mycoplasma genitalium* or *Mycoplasma suis*, are not in the list of predicted intracellulars. The two mentioned genomes only contain 36 and 37 marker COGs, respectively. Due to the strict filtering for complete genomes, these species were not considered. Predicting intracellular lifestyle in eggNOG 4.0 without applying the filter step yields 183 predicted intracellular bacteria (data not shown). For these 23 additional species, the reliability of the method is unclear, because of the inconsistent workflow. Future optimization of the completeness filter might reduce the number of missed intracellular bacteria.

Applying the model to distant relatives

In addition to the experiments above, also archaeal species in eggNOG 4.0 fulfilling the completeness criterion were classified based on the obligate intracellular model. Nine species, including Crenarchaeota from the class of *Thermoprotei*, were predicted as obligate intracellular, but in fact represent free-living microbes (Table 3.6). This misclassification is not surprising as metabolic pathways show pronounced differences between bacteria and archaea, with archaea lacking many classical pathways and containing unique modified variants [87]. A model based on a training set with exclusively intracellular bacteria is thus not suited for the analysis of archaeal genomes. This limitation can not be circumvented at the moment, due to a lack of known intracellular archaeal species. *Nanoarchaeum equitans* is the only currently known representative for host-dependent archaea.

Future models might be trained, however, for microbial eukaryotes. To this end, I conducted preliminary experiments. The model trained solely on bacteria was used to predict intracellular eukaryotic species in eggNOG 4.0. Several *Plasmodii, Leishmaniae, Trypanosomae, Encephalitozoa* as well as a *Toxoplasma* species are correctly predicted intracellular (without distinction of obligate and facultative manifestation). For others, prediction correctness is less clear. For example, one *Entamoeba* species was predicted intracellular. This genus contains internal parasites, but I could not find indications for intracellular lifestyle in the literature. All predicted intracellular eukaryotes are listed in Table 3.7. These results are promising for the applicability of phenotype prediction tools in eukaryotic microbes.

taxid	taxname
84599	Staphylothermus hellenicus
2280	Staphylothermus marinus
477693	Desulfurococcus kamchatkensis
2275	Desulfurococcus mucosus
54254	Thermosphaera aggregans
54248	Hyperthermus butylicus
2269	Thermofilum pendens
242703	Acidilobus saccharovorans
274854	uncultured marine group II euryarchaeote

Table 3.6: Predicted obligate intracellular archaea in eggNOG 4.0

taxid	taxname	taxid	taxname
5821	Plasmodium berghei	5664	Leishmania major
73239	Plasmodium yoelii yoelii	5660	Leishmania braziliensis
5825	Plasmodium chabaudi	5693	Trypanosoma cruzi
5850	Plasmodium knowlesi	5691	Trypanosoma brucei
5855	Plasmodium vivax	27973	Encephalitozoon hellem
5833	Plasmodium falciparum	6035	Encephalitozoon cuniculi
5875	Theileria parva	58839	Encephalitozoon intestinalis
5874	Theileria annulata	5911	Tetrahymena thermophila
29176	Neospora caninum	5888	Paramecium tetraurelia
5811	Toxoplasma gondii	5759	Entamoeba histolytica
5808	Cryptosporidium muris	46681	Entamoeba dispar
5807	Cryptosporidium parvum	5722	Trichomonas vaginalis
5661	Leishmania donovani	5762	Naegleria gruberi
5671	Leishmania infantum	31276	Perkinsus marinus
5665	Leishmania mexicana	5741	Giardia intestinalis

Table 3.7: Predicted obligate intracellular microbial eukaryotes in eggNOG 4.0

3.3 Phenotype Prediction in Metagenomics

The rapidly growing number of almost complete genome sequences extracted from environmental samples requires computational workflows for biological interpretation. Phenotype prediction in metagenomic context is, therefore, one major goal of the methodology presented in this thesis. I demonstrated well-performing phenotype prediction based on incomplete genome sequences in Section 3.1.4. In the following I present (to my knowledge) the first phenotype predictions based solely on sequences from metagenomic bins. All nineteen traits described in the previous chapters were predicted for high-quality bins from three different metagenomes:

- Coral (*Montipora hispida*) in the context of black band disease (BBD)

- Sponge (*Ianthella basta*) as a model system for symbiosis

- Biogas-fermenter (Northern Germany)

The results are provided on pages 73 till 75. CheckM [65] was used to determine genome completeness. Bins that are assumed to be less than 70 percent complete can not be classified with all models reliably (see Fig. 3.11, p. 53). They are indicated in the results table by a light gray font color. Predictions for these bins are to be interpreted cautiously.

3.3.1 Coral metagenome

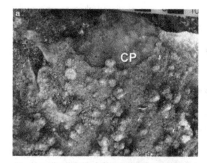

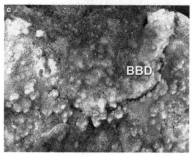

Figure 3.22: Transition of the disease from cyanobacterial patches via an intermediate state to black band disease. Reprinted by permission from Macmillan Publishers Ltd: ISME Journal [88], copyright 2010.

The coral *Montipora hispida* is investigated in the context of black band disease (BBD). This disease is caused by a mat-forming consortium of bacteria, archaea and eukaryotic microbes, which cause typical darkly colored lesions of coral tissue. Cyanobacteria dominate BBD consortia, but many other species like sulfur-cycle-associated bacteria, euryarchaeota and marine fungi are present as well. As the mat migrates across colonies, it causes necrotic tissue damage. Virulent infection of this type have been observed worldwide. BBD is acknowledged as a driving force for the loss of reef-building coral populations, and its effects are predicted to be aggravated by increasing water temperatures caused by the global climate change. BBD is often preceded by less virulent cyanobacterial patches (CP). As the disease progresses from CP to BBD (as depicted in Fig. 3.22), the microenvironment increasingly changes to anoxic, sulfide-rich conditions, which are important factors of BBD pathogenicity. Investigation of the consortium composition and its development might thus aid effective disease control and coral reef conservation [88] [89].

In the coral metagenome, there are seven bins that satisfy the completeness criterion. All of them are predicted to be aerobes, except for the facultative anaerobe bbd.Cya2. which is in alignment with the anoxic conditions in BBD. The latter is also predicted phototrophic and most likely represents a cyanobacterium. Another putative *Cyanobacterium* bin, cp.Cya1, is predicted phototrophic, but is only 60 percent complete. In this regime, the corresponding model is generally not reliable. Three near-complete bins are predicted halophiles, however, this model has rather poor accuracy overall. All bins are predicted to correspond to motile, gram-negative microbes. None of the metabolic or other features are predicted in any bin, except for combined.Alt1, which shows a signal for intracellular lifestyle. Considering this bin's completeness of around 50 percent and the fact, that intracellulars are mainly predicted by gene absence (see Subsection 3.2.3), this result is likely to be an artifact.

3.3.2   Sponge metagenome

Marine sponges are very simple metazoans that lack circulatory or digestive systems, but require constant water flow through their body for nutrient uptake. These sessile filter feeders partake in important marine biogeochemical processes and often live in symbiosis with microbial communities. Many of their functional roles are still unclear [90]. Nitrogen cycling in sponge holobionts has been investigated in several studies. Both proteobacterial and thaumarchaeotal ammonia oxidizers have been found associated with sponges.

The elephant ear sponge (*Ianthella basta*) belongs to the large class of *Demospongiae* within *Porifera*. It can be found throughout the Indo-Pacific region. *I. basta* from the Great Barrier Reef in Australia is used within the Department of Microbiology and Ecosystem Science as a marine model system and is investigated by meta-omics techniques. Compared to other sponges the diversity of its microbial symbionts is relatively low, making it a rather well-defined system. From a metagenomic perspective this facilitates precise binning.

The sponge metagenome exhibits similar features as the coral metagenome (see page 73). It consists of five near-complete bins and three others. All bins are predicted aerobes, except for the facultative anaerobic *Cyanobacteria* bin, which is only approximately 25 percent complete. Also in the sponge metagenome, all bins are predicted motile gram-negatives. Three obligate intracellular predictions correspond to three bins, which cover only small percentages of complete genomes. Two putative *Thaumarchaeota* bins are predicted nitrifiers (ammonia oxidizing archaea), which is in accordance with the results as described in Subsection 3.2.2. These two archaea are also predicted halophiles.

3.3.3 Biogas-fermenter metagenome

Biogas is a candidate for replacement of non-renewable energy sources like fossil fuels. It is a mixture of gases dominated by methane (CH_4) and carbon dioxide (CO_2). Biogas may also contain hydrogen (H_2), hydrogen sulfide (H_2S), nitrogen (N_2), ammonia (NH_3) and carbon monoxide (CO). Biomethane can fuel natural gas-powered devices (stoves, hot-water boilers, automobiles etc.) or can be used in combined cycle power plants in order to generate electricity as well as heat for teleheating. Biogas is produced from organic waste by microbial communities under anaerobic conditions. A variety of bacterial and archaeal microorganisms may participate in enzymatic degradation of complex organic compounds and perform steps from hydrolysis to methanogenesis. Knowledge about community composition, relative abundances and functional roles of individual species is necessary to optimize the process. Also abiotic factors influence the activity and performance of biogas-fermentation. These include substrate composition, temperature, pH value, mixing procedures and general geometry of the bioreactor. A sample was taken from a biogas-fermenter in northern Germany:

- Type: agricultural biogas plant

- Electric capacity: 536 kW

- Fermenter size: 2.800 m^3

- Fermentation conditions: 40 °C and pH 8

- Fermentation substrates:
  maize silage (69 %), cow manure (19 %), chicken manure (12 %)

The biogas-fermenter metagenome was analyzed for 51 high-quality bins, with completeness values above 80 percent and less than ten percent contamination. This anthropogenic environment has distinct characteristics compared to the two aquatic environments described in the previous subsections. A vast majority of bins is predicted anaerobic with the remaining not being labeled with respect to oxygen at all. This is in good alignment with fermentation being an anaerobic process as described above. Assuming that in fact all bins correspond to an anaerobic species, the anaerobic model would thus be 87 % accurate which perfectly matches the predicted accuracy based on cross-validation (compare Fig. 3.11, p. 53). Nearly 40 percent of all bins are predicted gram-negatives, 60 percent are classified as motiles. The endospore-forming and the thermophilic trait are predicted in 22 and 24 percent of all cases, respectively, and are indeed highly correlated (Pearson's $r = 0.72$). 20 percent of all bins are predicted obligate intracellulars. No predicted phototrophs, psychrophiles or nitrifiers are present in this metagenome. Moreover, no methanotrophs are predicted among the 51 bins. This result is of high interest, because methanotrophs could interfere with the energy harvest in the bioreactor. We can not, however, completely rule out the possibility of methanotrophs in this habitat, because the low-quality bins indicate presence of additional species, that could not be analyzed with this workflow.

		Coral meta-genome									
	Bin	1	2	3	4	5	6	7	8	9	10
CheckM measures	COMPLET	99,33	94,78	51,51	81,06	60,98	98,71	60,20	96,22	86,58	94,91
	CONTAMI	0,22	1,66	8,35	3,88	4,38	18,32	4,28	3,84	7,37	2,72
	STR-HET	0,00	0,00	45,21	35,71	63,64	12,50	72,00	31,82	28,00	46,67
Phenotype predictions	AEROBE		X		X	X	X	X	X	X	X
	ANAER										
	FAC-AN	X									
	GRAMNEG	X	X	X	X	X	X	X	X	X	X
	HALO		X	X	X	X					
	MOTILE	X	X	X	X	X	X	X	X	X	X
	PHOTO	X						X			
	PSYCHO										
	SPORE										
	THERM										
	METHANO										
	INTRA			X							
	OBL.INT										
	FAC.INT										
	NITR										
	AOA										
	AOB										
	AMM-OX										
	NITR-OX										

Abbreviations

COMPLET	Completeness
CONTAMI	Contamination
STR-HET	Strain heterogeneity
AEROBE	aerobic
ANAER	anaerobic
FAC-AN	facultative anaerobic
GRAMNEG	gram negative
HALO	halophilic
MOTILE	motile
PHOTO	phototrophic
PSYCHO	psychrophilic
SPORE	Endospore-forming
THERM	thermophilic
METHANO	methanotropic
INTRA	intracellular
OBL.INT	obligate intracellular
FAC.INT	facultative intracellular
NITR	nitrification
AOA	Amm-ox archaeon
AOB	Amm-ox bacterium
AMM-OX	ammonium oxidizing
NITR-OX	nitrite oxidizing

Bins

1	bbd.Cya2
2	combined.Alpha3
3	combined.Alt1
4	combined.Alt2
5	combined.Fla1
6	combined.Oce
7	cp.Cya1
8	cp.Cyt1
9	cp.Cyt2
10	cp.Cyt3

		Sponge meta-genome							
	Bin	1	2	3	4	5	6	7	8
CheckM measures	COMPLET	79,55	26,68	87,12	95,10	95,72	95,72	11,33	0,00
	CONTAMI	0,26	1,48	67,70	1,72	16,18	2,43	0,49	0,00
	STR-HET	0,00	43,75	90,00	50,00	86,67	100	100	0,00
Phenotype predictions	AEROBE	X		X	X	X	X		
	ANAER								
	FAC-AN		X						
	GRAMNEG	X	X	X	X	X	X	X	X
	HALO					X	X		
	MOTILE	X	X	X	X	X	X	X	X
	PHOTO								
	PSYCHO								
	SPORE								
	THERM								
	METHANO								
	INTRA		X					X	X
	OBL.INT		X					X	X
	FAC.INT								
	NITR					X	X		
	AOA					X	X		
	AOB								
	AMM-OX					X	X		
	NITR-OX								

Abbreviations

COMPLET	Completeness
CONTAMI	Contamination
STR-HET	Strain heterogeneity
AEROBE	aerobic
ANAER	anaerobic
FAC-AN	facultative anaerobic
GRAMNEG	gram negative
HALO	halophilic
MOTILE	motile
PHOTO	phototrophic
PSYCHO	psychrophilic
SPORE	Endospore-forming
THERM	thermophilic
METHANO	methanotropic
INTRA	intracellular
OBL.INT	obligate intracellular
FAC.INT	facultative intracellular
NITR	nitrification
AOA	Amm-ox archaeon
AOB	Amm-ox bacterium
AMM-OX	ammonium oxidizing
NITR-OX	nitrite oxidizing

Bins

1	Alphaproteobacteria
2	Cyanobacteria
3	Gammaproteobacteria
4	Planctomycetes
5	Thaumarchaeota_c
6	Thaumarchaeota_cnw
7	Thaumarchaeota_n
8	Thaumarchaeota_w

	Bin	Biogas-fermenter meta-genome (part 1)									
		2	3	30	31	40	47	65	69	70	80
CheckM measures	COMPLET	82,22	96,61	87,12	99,3	95,11	90,03	98,6	96,24	91,13	95,1
	CONTAMI	5,20	2,97	0,99	7,23	0,13	3,41	7,69	3,73	5,2	2,97
	STR-HET	58,33	60	80	100	100	57,14	100	90,91	72,73	50
Phenotype predictions	AEROBE										
	ANAER	X	X	X	X	X		X	X	X	X
	FAC-AN										
	GRAMNEG	X		X	X		X	X			
	HALO										
	MOTILE		X	X	X						X
	PHOTO										
	PSYCHO										
	SPORE		X							X	X
	THERM		X							X	X
	METHANO										
	INTRA						X		X		
	OBL.INT						X		X		
	FAC.INT										
	NITR										
	AOA										
	AOB										
	AMM-OX										
	NITR-OX										

	Bin	84	85	88	90	97	121	122	162	172	173
CheckM measures	COMPLET	92,88	99,84	96,24	99,3	97,09	98,84	97,74	88,36	96,24	87,77
	CONTAMI	1,9	0,03	1,08	9,91	1,43	4,4	2,85	7,07	2,07	4,39
	STR-HET	71,43	100	100	96,3	90	100	75	70	75	100
Phenotype predictions	AEROBE										
	ANAER		X	X	X	X		X	X	X	
	FAC-AN										
	GRAMNEG		X	X	X		X	X		X	X
	HALO		X								
	MOTILE	X			X		X			X	X
	PHOTO										
	PSYCHO										
	SPORE	X									
	THERM										
	METHANO										
	INTRA							X		X	X
	OBL.INT							X		X	X
	FAC.INT										
	NITR										
	AOA										
	AOB										
	AMM-OX										
	NITR-OX										

	Bin	Biogas-fermenter meta-genome (part 2)									
		177	185	190	199	205	207	212	215	237	243
CheckM measures	COMPLET	88,45	84,97	98,28	99,94	94,38	87,5	98,31	100	90,03	89,14
	CONTAMI	3,06	4,18	6,74	7,69	4,2	0,5	6,78	8,77	2,82	3,51
	STR-HET	60	80	91,3	60	85,71	50	100	100	50	50
Phenotype predictions	AEROBE										
	ANAER	X	X	X	X	X	X	X	X	X	X
	FAC-AN										
	GRAMNEG			X	X			X	X		X
	HALO										
	MOTILE	X		X	X	X		X	X		
	PHOTO										
	PSYCHO										
	SPORE	X		X					X		
	THERM	X		X				X	X		
	METHANO										
	INTRA										
	OBL.INT										
	FAC.INT										
	NITR										
	AOA										
	AOB										
	AMM-OX										
	NITR-OX										

Abbreviations

COMPLET Completeness
CONTAMI Contamination
STR-HET Strain heterogeneity

AEROBE aerobic
ANAER anaerobic
FAC-AN facultative anaerobic
GRAMNEG gram negative
HALO halophilic
MOTILE motile
PHOTO phototrophic
PSYCHO psychrophilic
SPORE Endospore-forming
THERM thermophilic
METHANO methanotropic
INTRA intracellular
OBL.INT obligate intracellular
FAC.INT facultative intracellular
NITR nitrification
AOA Amm-ox archaeon
AOB Amm-ox bacterium
AMM-OX ammonium oxidizing
NITR-OX nitrite oxidizing

Bins
(numbers equal original bin names)

	Bin	Biogas-fermenter meta-genome (part 3)									
		246	108-3	166-2	186-2	192-1	206-2	233-1	233-3	235-1	235-2
CheckM measures	COMPLET	85,81	88,35	95,48	90,72	96,77	81,97	93,96	84,14	86,35	93,17
	CONTAMI	5,09	5,23	7,14	2,06	1,61	5,41	1,68	4,49	4,36	4,79
	STR-HET	78,57	55,56	52,38	100	50	72,22	75	80	70	50
Phenotype predictions	AEROBE										
	ANAER	X	X	X	X	X	X	X	X	X	X
	FAC-AN										
	GRAMNEG			X	X			X			
	HALO										
	MOTILE	X	X	X		X	X		X		X
	PHOTO										
	PSYCHO										
	SPORE		X				X				
	THERM		X	X			X				
	METHANO										
	INTRA									X	
	OBL.INT									X	
	FAC.INT										
	NITR										
	AOA										
	AOB										
	AMM-OX										
	NITR-OX										

	Bin	35-1	35-2	38-1	6-1	6-2	60-2	61-1	78-1	93-2	15-1
CheckM measures	COMPLET	91,05	98,3	92,06	96	88,55	92,58	93,1	94,07	97,92	98,31
	CONTAMI	1,34	8,58	6,33	1,33	1,82	2,54	0,67	8,47	5,17	5,93
	STR-HET	66,67	92,59	100	100	60	50	100	62,5	67,74	75
Phenotype predictions	AEROBE										
	ANAER	X	X	X				X	X	X	X
	FAC-AN										
	GRAMNEG										
	HALO										
	MOTILE	X	X	X				X	X	X	X
	PHOTO										
	PSYCHO										
	SPORE									X	X
	THERM							X			X
	METHANO										
	INTRA				X	X	X	X			
	OBL.INT				X	X	X	X			
	FAC.INT										
	NITR										
	AOA										
	AOB										
	AMM-OX										
	NITR-OX										

	Bin	76-1					
CheckM measures		COMPLET	96,55	CONTAMI	8,62	STR-HET	100
Phenotype predictions	ANAER	X					

Chapter 4

Conclusion and Outlook

Inspired by the rapid progress in metagenomics, producing hundreds of high-quality genome bins from even a single modern study (e.g., [5]), I have explored how phenotypic trait prediction might better contribute to microbiology and microbial ecology. I have therefore put particular emphasis on incomplete genomes and vastly increasing data amounts. I could demonstrate the stability of the predictive power for phenotypic traits by reproducing earlier results, indicating that this method is not perturbed by the rapid growth of genome databases. A new software tool was developed that facilitates the in-depth analysis of phenotype models. It allows associating expected and unexpected protein functions with particular traits. Most of the traits can be reliably predicted in only 60-70% complete genomes, which allows reasonable predictions in genome bins from metagenomes. Testing bins from three different metagenomes indeed resulted in many predictions that are in perfect alignment with known environmental factors and gave no indication for methanotrophic microorganisms in a biogas-fermenter.

New models have been created for the prediction of methanotrophs and nitrifiers as proofs-of-principle, clearly demonstrating highly accurate prediction of simple, yet ecologically important metabolic traits, as soon as sufficient training data are available. Although these models recover known functional markers, they substantially extend the marker concept by associating many further genes to the phenotypic traits. I have also established a new phenotypic model that predicts intracellular microorganisms. Thereby I could demonstrate that also independently evolved phenotypic traits, characterized by genome reduction, can be reliably predicted based on comparative genomics. This model is an example of a trait that can not be associated to single functional marker genes. The predictive power of its model therefore arises from the combination of multiple (mainly absence) genotypic signals. Currently ongoing work in the group indicates good performance of phenotype prediction also for other traits, such as bacterial secretion systems, which are of high interest in the research of host-pathogen interactions. For all the traits mentioned here, predictions in COG genotype space turned out to work well, with the highest level COGs being superior to lower level NOGs, as has been shown for bactNOGs. My colleague Valerie Eichinger, who works on the secretion systems, performed predictions in different genotype spaces, like e.g., Pfam protein domains and families, but found these to be inferior to COG-based predictions. Future predictions will thus be performed in COG space.

The results presented in this thesis suggest that the extended PICA framework can be used to automatically annotate phenotypes in near-complete microbial genome sequences, as generated in large numbers by modern metagenomics. The GenTraitor pipeline will also prove useful for annotating phenotypes in complete microbial genomes that are being deposited in public databases, but not described in detail in any scientific article. Only automated bioinformatic methodologies will be able to keep the pace with the large amounts of generated data.

In the future, technical improvements may be implemented in the PICA framework. Current models are based on SVMs with linear kernels, which are trained on very few datapoints compared to the number of features. The LIBLINEAR package [91] for linear SVMs is a sister project to LIBSVM. It uses different optimization routines, which are highly efficient on large, sparse datasets. Thus, the development of a LIBLINEAR plug-in for PICA would most likely further improve its computational performance. Phenotype prediction itself is, however, not the rate-limiting step in the process of phenotype annotation. Currently, mapping protein sequences of new genomes to orthologous groups accounts for the largest part of computational cost. Improvements in this process would therefore be highly beneficial to phenotype annotation. During the training phase of a new model, another step is much more time-consuming: Obtaining phenotype target labels for training data currently relies mainly on expert knowledge and literature mining. Comprehensive phenotype databases would mitigate this problem. If such databases are not available, automatic knowledge extraction from literature using natural language processing techniques might facilitate the creation of many more models, covering larger grounds of biological and ecological functions.

Abbreviations

API application programming interface. 32, 37

ARM association rule mining. 2, 3, 12, 14, 15, 32, 35, 44–46

BBD black band disease. 68–70

BBH bidirectional best hit. 22, 25

COG Cluster of Orthologous Groups. 2, 3, 15, 22–26, 32, 37–39, 41, 48–52, 64–66, 78

CPAR classification based on predictive association rules. 3, 32, 33, 44, 45

CWMI conditionally weighted mutual information. 3, 33, 34

LUCA last universal common ancestor. 2, 48–51

MI mutual information. 30, 33, 34

NOG Non-supervised Orthologous Groups. 2, 15, 22, 26, 48, 50, 78

PICA Phenotype Investigation with Classification Algorithms. 3, 27, 28, 32, 33, 35, 37–39, 44–46, 48, 51, 78

RBF radial basis function. 21, 46

SVC support vector classification. 3, 13, 35

SVM support vector machine. XI, 12, 16–19, 22, 35–38, 45, 46, 53, 78

References

[1] R. I. Amann, W. Ludwig, and K.-H. Schleifer, "Phylogenetic identification and in situ detection of individual microbial cells without cultivation.," *Microbiological reviews*, vol. 59,
no. 1, pp. 143–169, 1995.

[2] E. A. Franzosa, T. Hsu, A. Sirota-Madi, A. Shafquat, G. Abu-Ali, X. C. Morgan, and
C. Huttenhower, "Sequencing and beyond: integrating molecular'omics' for microbial
community profiling," *Nature Reviews Microbiology*, 2015.

[3] S. J. Callister, M. J. Wilkins, C. D. Nicora, K. H. Williams, J. F. Banfield, N. C. Ver-
Berkmoes, R. L. Hettich, L. Nï£¡Guessan, P. J. Mouser, H. Elifantz, *et al.*, "Analysis
of biostimulated microbial communities from two field experiments reveals temporal and
spatial differences in proteome profiles," *Environmental science & technology*, vol. 44,
no. 23, pp. 8897–8903, 2010.

[4] M. Albertsen, P. Hugenholtz, A. Skarshewski, K. L. Nielsen, G. W. Tyson, and P. H.
Nielsen, "Genome sequences of rare, uncultured bacteria obtained by differential coverage
binning of multiple metagenomes," *Nature biotechnology*, vol. 31, no. 6, pp. 533–538,
2013.

[5] C. T. Brown, L. A. Hug, B. C. Thomas, I. Sharon, C. J. Castelle, A. Singh, M. J.
Wilkins, K. C. Wrighton, K. H. Williams, and J. F. Banfield, "Unusual biology across a
group comprising more than 15% of domain bacteria.," *Nature*, vol. 523, pp. 208–211,
Jul 2015.

[6] A. M. Altenhoff, N. Škunca, N. Glover, C.-M. Train, A. Sueki, I. Piližota, K. Gori,
B. Tomiczek, S. Müller, H. Redestig, *et al.*, "The oma orthology database in 2015:

function predictions, better plant support, synteny view and other improvements," *Nucleic acids research*, p. gku1158, 2014.

[7] S. Powell, K. Forslund, D. Szklarczyk, K. Trachana, A. Roth, J. Huerta-Cepas, T. Gabaldón, T. Rattei, C. Creevey, M. Kuhn, L. J. Jensen, C. von Mering, and P. Bork, "eggnog v4.0: nested orthology inference across 3686 organisms," *Nucleic Acids Research*, vol. 42, no. D1, pp. D231–D239, 2014.

[8] M. Y. Galperin, K. S. Makarova, Y. I. Wolf, and E. V. Koonin, "Expanded microbial genome coverage and improved protein family annotation in the cog database," *Nucleic acids research*, p. gku1223, 2014.

[9] M. Kanehisa, S. Goto, Y. Sato, M. Kawashima, M. Furumichi, and M. Tanabe, "Data, information, knowledge and principle: back to metabolism in kegg," *Nucleic acids research*, vol. 42, no. D1, pp. D199–D205, 2014.

[10] D. Szklarczyk, A. Franceschini, S. Wyder, K. Forslund, D. Heller, J. Huerta-Cepas, M. Simonovic, A. Roth, A. Santos, K. P. Tsafou, *et al.*, "String v10: protein–protein interaction networks, integrated over the tree of life," *Nucleic acids research*, p. gku1003, 2014.

[11] T. Tatusova, S. Ciufo, S. Federhen, B. Fedorov, R. McVeigh, K. O'Neill, I. Tolstoy, and L. Zaslavsky, "Update on refseq microbial genomes resources," *Nucleic acids research*, p. gku1062, 2014.

[12] E. V. Koonin, "Evolution of genome architecture," *The international journal of biochemistry & cell biology*, vol. 41, no. 2, pp. 298–306, 2009.

[13] M. C. Chibucos, A. E. Zweifel, J. C. Herrera, W. Meza, S. Eslamfam, P. Uetz, D. A. Siegele, J. C. Hu, and M. G. Giglio, "An ontology for microbial phenotypes," *BMC microbiology*, vol. 14, no. 1, p. 294, 2014.

[14] D. Medini, C. Donati, H. Tettelin, V. Masignani, and R. Rappuoli, "The microbial pangenome," *Current opinion in genetics & development*, vol. 15, no. 6, pp. 589–594, 2005.

[15] M. J. Kampschreur, R. Kleerebezem, C. Picioreanu, L. Bakken, L. Bergaust, S. de Vries, M. S. Jetten, and M. C. van Loosdrecht, "Metabolic modeling of denitrification in agrobacterium tumefaciens: a tool to study inhibiting and activating compounds for the denitrification pathway," *Frontiers in microbiology*, vol. 3, 2012.

[16] J.-H. Rotthauwe, K.-P. Witzel, and W. Liesack, "The ammonia monooxygenase structural gene amoa as a functional marker: molecular fine-scale analysis of natural ammonia-oxidizing populations.," *Applied and environmental microbiology*, vol. 63, no. 12, pp. 4704–4712, 1997.

[17] M. Tamura and P. D'haeseleer, "Microbial genotype–phenotype mapping by class association rule mining," *Bioinformatics*, vol. 24, no. 13, pp. 1523–1529, 2008.

[18] N. J. MacDonald and R. G. Beiko, "Efficient learning of microbial genotype–phenotype association rules," *Bioinformatics*, vol. 26, no. 15, pp. 1834–1840, 2010.

[19] C.-C. Chang and C.-J. Lin, "Libsvm: A library for support vector machines," *ACM Transactions on Intelligent Systems and Technology (TIST)*, vol. 2, no. 3, p. 27, 2011.

[20] J. Reece, L. A. Urry, M. L. Cain, S. A. Wasserman, P. V. Minorsky, and R. B. Jackson, *Campbell biology*. Pearson Higher Education AU, 10 ed., 2014.

[21] Y. Ogura, D. K. Bonen, N. Inohara, D. L. Nicolae, F. F. Chen, R. Ramos, H. Britton, T. Moran, R. Karaliuskas, R. H. Duerr, *et al.*, "A frameshift mutation in nod2 associated with susceptibility to crohn's disease," *Nature*, vol. 411, no. 6837, pp. 603–606, 2001.

[22] P. A. Zimmerman, A. Buckler-White, G. Alkhatib, T. Spalding, J. Kubofcik, C. Combadiere, D. Weissman, O. Cohen, A. Rubbert, G. Lam, *et al.*, "Inherited resistance to hiv-1 conferred by an inactivating mutation in cc chemokine receptor 5: studies in populations with contrasting clinical phenotypes, defined racial background, and quantified risk.," *Molecular Medicine*, vol. 3, no. 1, p. 23, 1997.

[23] M. Wagner, A. Loy, M. Klein, N. Lee, N. B. Ramsing, D. A. Stahl, and M. W. Friedrich, "Functional marker genes for identification of sulfate-reducing prokaryotes," *Methods in enzymology*, vol. 397, pp. 469–489, 2005.

[24] H. Shen, "Neuroscience: The hard science of oxytocin.," *Nature*, vol. 522, no. 7557, pp. 410–412, 2015.

[25] M. T. Madigan, D. P. Clark, D. Stahl, and J. M. Martinko, *Brock Biology of Microorganisms 13th edition*. Benjamin Cummings, 2010.

[26] J. C. Wooley, A. Godzik, and I. Friedberg, "A primer on metagenomics," *PLoS Comput Biol*, vol. 6, no. 2, p. e1000667, 2010.

[27] R. J. Klein, C. Zeiss, E. Y. Chew, J.-Y. Tsai, R. S. Sackler, C. Haynes, A. K. Henning, J. P. SanGiovanni, S. M. Mane, S. T. Mayne, *et al.*, "Complement factor h polymorphism in age-related macular degeneration," *Science*, vol. 308, no. 5720, pp. 385–389, 2005.

[28] T. Fall and E. Ingelsson, "Genome-wide association studies of obesity and metabolic syndrome," *Molecular and cellular endocrinology*, vol. 382, no. 1, pp. 740–757, 2014.

[29] C. Sandholt, T. Hansen, and O. Pedersen, "Beyond the fourth wave of genome-wide obesity association studies," *Nutrition & diabetes*, vol. 2, no. 7, p. e37, 2012.

[30] J. Couzin-Frankel, "Major heart disease genes prove elusive," *Science*, vol. 328, no. 5983, pp. 1220–1221, 2010.

[31] H. Marchandin, C. Teyssier, J. Campos, H. Jean-Pierre, F. Roger, B. Gay, J.-P. Carlier, and E. Jumas-Bilak, "Negativicoccus succinicivorans gen. nov., sp. nov., isolated from human clinical samples, emended description of the family veillonellaceae and description of negativicutes classis nov., selenomonadales ord. nov. and acidaminococcaceae fam. nov. in the bacterial phylum firmicutes," *International journal of systematic and evolutionary microbiology*, vol. 60, no. 6, pp. 1271–1279, 2010.

[32] P. J. Turnbaugh, R. E. Ley, M. A. Mahowald, V. Magrini, E. R. Mardis, and J. I. Gordon, "An obesity-associated gut microbiome with increased capacity for energy harvest," *Nature*, vol. 444, no. 7122, pp. 1027–131, 2006.

[33] M. M. Finucane, T. J. Sharpton, T. J. Laurent, and K. S. Pollard, "A taxonomic signature

of obesity in the microbiome? getting to the guts of the matter," *PLoS ONE*, vol. 9, p. e84689, 01 2014.

[34] The Editors of The Encyclopædia Britannica, "Learning." http://www.britannica. com/topic/learning. Accessed: 2015-08-05.

[35] D. Kumaran, J. J. Summerfield, D. Hassabis, and E. A. Maguire, "Tracking the emergence of conceptual knowledge during human decision making," *Neuron*, vol. 63, no. 6, pp. 889–901, 2009.

[36] B.-J. Yoon, "Hidden markov models and their applications in biological sequence analysis," *Current genomics*, vol. 10, no. 6, p. 402, 2009.

[37] K. Tokuda, Y. Nankaku, T. Toda, H. Zen, J. Yamagishi, and K. Oura, "Speech synthesis based on hidden markov models," *Proceedings of the IEEE*, vol. 101, no. 5, pp. 1234–1252, 2013.

[38] C. Karlof and D. Wagner, *Hidden Markov model cryptanalysis*. Springer, 2003.

[39] A. Narayanan and V. Shmatikov, "Fast dictionary attacks on passwords using time-space tradeoff," in *Proceedings of the 12th ACM conference on Computer and communications security*, pp. 364–372, ACM, 2005.

[40] Y. Taigman, M. Yang, M. Ranzato, and L. Wolf, "Deepface: Closing the gap to human-level performance in face verification," in *Computer Vision and Pattern Recognition (CVPR), 2014 IEEE Conference on*, pp. 1701–1708, IEEE, 2014.

[41] A. Cully, J. Clune, D. Tarapore, and J.-B. Mouret, "Robots that can adapt like animals," *Nature*, vol. 521, no. 7553, pp. 503–507, 2015.

[42] S. Marsland, *Machine learning: an algorithmic perspective*. CRC press, 1 ed., 2009.

[43] C. M. Bishop, *Pattern recognition and machine learning*. springer, 2006.

[44] S. Marsland, *Machine learning: an algorithmic perspective*. CRC press, 2 ed., 2014.

[45] I. Artamonova, S. Kramer, and D. Frishman, "Data mining in genome annotation," in *Modern Genome Annotation*, pp. 191–212, Springer, 2008.

[46] R. Agrawal, T. Imieliński, and A. Swami, "Mining association rules between sets of items in large databases," in *ACM SIGMOD Record*, vol. 22, pp. 207–216, ACM, 1993.

[47] C. Duhigg, "How companies learn your secrets," *The New York Times*, 2 2012.

[48] C. Zhang and S. Zhang, *Association rule mining: models and algorithms*. Springer-Verlag, 2002.

[49] I. I. Artamonova, G. Frishman, M. S. Gelfand, and D. Frishman, "Mining sequence annotation databanks for association patterns," *Bioinformatics*, vol. 21, no. Suppl 3, pp. iii49–iii57, 2005.

[50] Y. Zhang and J. C. Rajapakse, *Machine learning in bioinformatics*, vol. 4. John Wiley & Sons, 2009.

[51] R. Gutierrez-Osuna, "L2: Support vector machines," *Lecture Notes*, p. 12. http://research.cs.tamu.edu/prism/lectures/pr/pr_121.pdf. Retrieved on 2015-08-25.

[52] C. J. Lin, C.-W. Hsu, and C.-C. Chang, "A practical guide to support vector classification," *National Taiwan U.*, 2003.

[53] A. M. Altenhoff and C. Dessimoz, "Inferring orthology and paralogy," in *Evolutionary genomics*, pp. 259–279, Springer, 2012.

[54] R. L. Tatusov, E. V. Koonin, and D. J. Lipman, "A genomic perspective on protein families," *Science*, vol. 278, no. 5338, pp. 631–637, 1997.

[55] L. J. Jensen, P. Julien, M. Kuhn, C. von Mering, J. Muller, T. Doerks, and P. Bork, "eggnog: automated construction and annotation of orthologous groups of genes," *Nucleic acids research*, vol. 36, no. suppl 1, pp. D250–D254, 2008.

[56] J. Muller, D. Szklarczyk, P. Julien, I. Letunic, A. Roth, M. Kuhn, S. Powell, C. von Mering, T. Doerks, L. J. Jensen, and P. Bork, "eggnog v2.0: extending the evolutionary genealogy of genes with enhanced non-supervised orthologous groups, species and

functional annotations," *Nucleic Acids Research*, vol. 38, no. suppl 1, pp. D190–D195, 2010.

[57] D. Hyatt, G.-L. Chen, P. F. LoCascio, M. L. Land, F. W. Larimer, and L. J. Hauser, "Prodigal: prokaryotic gene recognition and translation initiation site identification," *BMC bioinformatics*, vol. 11, no. 1, p. 119, 2010.

[58] D. M. Kristensen, L. Kannan, M. K. Coleman, Y. I. Wolf, A. Sorokin, E. V. Koonin, and A. Mushegian, "A low-polynomial algorithm for assembling clusters of orthologous groups from intergenomic symmetric best matches," *Bioinformatics*, vol. 26, no. 12, pp. 1481–1487, 2010.

[59] V. M. Markowitz, I.-M. A. Chen, K. Palaniappan, K. Chu, E. Szeto, M. Pillay, A. Ratner, J. Huang, T. Woyke, M. Huntemann, *et al.*, "Img 4 version of the integrated microbial genomes comparative analysis system," *Nucleic acids research*, p. gkt963, 2013.

[60] B. D. Ondov, N. H. Bergman, and A. M. Phillippy, "Interactive metagenomic visualization in a web browser," *BMC bioinformatics*, vol. 12, no. 1, p. 385, 2011.

[61] X. Yin and J. Han, "Cpar: Classification based on predictive association rules.," in *SDM*, vol. 3, pp. 369–376, SIAM, 2003.

[62] Y.-W. Chang and C.-J. Lin, "Feature ranking using linear svm," *Causation and Prediction Challenge Challenges in Machine Learning*, vol. 2, p. 47, 2008.

[63] S. F. Altschul, T. L. Madden, A. A. Schäffer, J. Zhang, Z. Zhang, W. Miller, and D. J. Lipman, "Gapped blast and psi-blast: a new generation of protein database search programs," *Nucleic acids research*, vol. 25, no. 17, pp. 3389–3402, 1997.

[64] R. D. Finn, A. Bateman, J. Clements, P. Coggill, R. Y. Eberhardt, S. R. Eddy, A. Heger, K. Hetherington, L. Holm, J. Mistry, *et al.*, "Pfam: the protein families database," *Nucleic acids research*, p. gkt1223, 2013.

[65] D. H. Parks, M. Imelfort, C. T. Skennerton, P. Hugenholtz, and G. W. Tyson, "Checkm:

assessing the quality of microbial genomes recovered from isolates, single cells, and meta-genomes," *Genome research*, pp. gr–186072, 2015.

[66] M. d. l. M. Camara, L. A. Bouvier, M. R. Miranda, and C. A. Pereira, "The flagellar adenylate kinases of trypanosoma cruzi," *FEMS Microbiology Letters*, vol. 362, no. 1, pp. 1–5, 2015.

[67] H. Zhang and D. R. Mitchell, "Cpc1, a chlamydomonas central pair protein with an adenylate kinase domain," *Journal of Cell Science*, vol. 117, no. 18, pp. 4179–4188, 2004.

[68] D. R. Mende, S. Sunagawa, G. Zeller, and P. Bork, "Accurate and universal delineation of prokaryotic species," *Nature methods*, vol. 10, no. 9, pp. 881–884, 2013.

[69] S. S. Keerthi and C.-J. Lin, "Asymptotic behaviors of support vector machines with gaussian kernel," *Neural computation*, vol. 15, no. 7, pp. 1667–1689, 2003.

[70] M. H. Sazinsky and S. J. Lippard, "Methane monooxygenase: Functionalizing methane at iron and copper," in *Sustaining Life on Planet Earth: Metalloenzymes Mastering Dioxygen and Other Chewy Gases*, pp. 205–256, Springer, 2015.

[71] I. R. McDonald and J. C. Murrell, "The particulate methane monooxygenase gene pmoa and its use as a functional gene probe for methanotrophs," *FEMS microbiology letters*, vol. 156, no. 2, pp. 205–210, 1997.

[72] K. F. Ettwig, M. K. Butler, D. Le Paslier, E. Pelletier, S. Mangenot, M. M. Kuypers, F. Schreiber, B. E. Dutilh, J. Zedelius, D. De Beer, *et al.*, "Nitrite-driven anaerobic methane oxidation by oxygenic bacteria," *Nature*, vol. 464, no. 7288, pp. 543–548, 2010.

[73] I. Tamas, S. N. Dedysh, W. Liesack, M. B. Stott, M. Alam, J. C. Murrell, and P. F. Dunfield, "Complete genome sequence of beijerinckia indica subsp. indica," *Journal of bacteriology*, vol. 192, no. 17, pp. 4532–4533, 2010.

[74] A. Fjellbirkeland, V. Torsvik, and L. Øvreås, "Methanotrophic diversity in an agricultural

soil as evaluated by denaturing gradient gel electrophoresis profiles of pmoa, mxaf and 16s rdna sequences," *Antonie van Leeuwenhoek*, vol. 79, no. 2, pp. 209–217, 2001.

[75] S. Vuilleumier, T. Nadalig, M. F. U. Haque, G. Magdelenat, A. Lajus, S. Roselli, E. E. Muller, C. Gruffaz, V. Barbe, C. Médigue, *et al.*, "Complete genome sequence of the chloromethane-degrading hyphomicrobium sp. strain mc1," *Journal of bacteriology*, vol. 193, no. 18, pp. 5035–5036, 2011.

[76] H. Koch, A. Galushko, M. Albertsen, A. Schintlmeister, C. Gruber-Dorninger, S. Lücker, E. Pelletier, D. Le Paslier, E. Spieck, A. Richter, *et al.*, "Growth of nitrite-oxidizing bacteria by aerobic hydrogen oxidation," *Science*, vol. 345, no. 6200, pp. 1052–1054, 2014.

[77] M. Pester, F. Maixner, D. Berry, T. Rattei, H. Koch, S. Lücker, B. Nowka, A. Richter, E. Spieck, E. Lebedeva, *et al.*, "Nxrb encoding the beta subunit of nitrite oxidoreductase as functional and phylogenetic marker for nitrite-oxidizing nitrospira," *Environmental microbiology*, vol. 16, no. 10, pp. 3055–3071, 2014.

[78] J. Schott, B. M. Griffin, and B. Schink, "Anaerobic phototrophic nitrite oxidation by thiocapsa sp. strain ks1 and rhodopseudomonas sp. strain lq17," *Microbiology*, vol. 156, no. 8, pp. 2428–2437, 2010.

[79] M. Wiederstein, M. Gruber, K. Frank, F. Melo, and M. J. Sippl, "Structure-based characterization of multiprotein complexes," *Structure*, vol. 22, no. 7, pp. 1063–1070, 2014.

[80] G. Greub and D. Raoult, "Microorganisms resistant to free-living amoebae," *Clinical microbiology reviews*, vol. 17, no. 2, pp. 413–433, 2004.

[81] S. Schmitz-Esser, E. R. Toenshoff, S. Haider, E. Heinz, V. M. Hoenninger, M. Wagner, and M. Horn, "Diversity of bacterial endosymbionts of environmental acanthamoeba isolates," *Applied and environmental microbiology*, vol. 74, no. 18, pp. 5822–5831, 2008.

[82] M. Horn, "Chlamydiae as symbionts in eukaryotes," *Annu. Rev. Microbiol.*, vol. 62, pp. 113–131, 2008.

[83] C. Toft and S. G. Andersson, "Evolutionary microbial genomics: insights into bacterial host adaptation," *Nature Reviews Genetics*, vol. 11, no. 7, pp. 465–475, 2010.

[84] F. Schulz and M. Horn, "Intranuclear bacteria: inside the cellular control center of eukaryotes," *Trends in cell biology*, vol. 25, no. 6, pp. 339–346, 2015.

[85] S. J. Giovannoni, H. J. Tripp, S. Givan, M. Podar, K. L. Vergin, D. Baptista, L. Bibbs, J. Eads, T. H. Richardson, M. Noordewier, *et al.*, "Genome streamlining in a cosmopolitan oceanic bacterium," *science*, vol. 309, no. 5738, pp. 1242–1245, 2005.

[86] C. M. Alsmark, A. C. Frank, E. O. Karlberg, B.-A. Legault, D. H. Ardell, B. Canbäck, A.-S. Eriksson, A. K. Näslund, S. A. Handley, M. Huvet, *et al.*, "The louse-borne human pathogen bartonella quintana is a genomic derivative of the zoonotic agent bartonella henselae," *Proceedings of the National Academy of Sciences of the United States of America*, vol. 101, no. 26, pp. 9716–9721, 2004.

[87] C. Bräsen, D. Esser, B. Rauch, and B. Siebers, "Carbohydrate metabolism in archaea: current insights into unusual enzymes and pathways and their regulation," *Microbiology and Molecular Biology Reviews*, vol. 78, no. 1, pp. 89–175, 2014.

[88] Y. Sato, B. L. Willis, and D. G. Bourne, "Successional changes in bacterial communities during the development of black band disease on the reef coral, montipora hispida," *The ISME journal*, vol. 4, no. 2, pp. 203–214, 2010.

[89] Y. Sato, B. L. Willis, and D. G. Bourne, "Pyrosequencing-based profiling of archaeal and bacterial 16s rrna genes identifies a novel archaeon associated with black band disease in corals," *Environmental microbiology*, vol. 15, no. 11, pp. 2994–3007, 2013.

[90] F. U. Moeller, N. S. Webster, F. Behnam, D. Domman, M. Albertsen, S. Markert, D. Turaev, D. Becher, T. Rattei, T. Schweder, A. Richter, P. H. Nielsen, and M. Wagner, "Metaproteogenomic and physiological characterization of a dominant marine sponge thaumarchaeal symbiont reveals mechanisms for complex organic matter cycling and functional convergence among sponge thaumarchaeotes," *in preparation*, 2015.

[91] R.-E. Fan, K.-W. Chang, C.-J. Hsieh, X.-R. Wang, and C.-J. Lin, "Liblinear: A library for large linear classification," *The Journal of Machine Learning Research*, vol. 9, pp. 1871–1874, 2008.

Appendix A

Source code

Listing A.1: Linear SVM feature ranking

```
"""
Extract the most important features of an SVM model, i.e. the COGs/NOGs with
the highest predicted impact on presence/absence of the phenotype.
This is done by calculating the primal variable w (weights vector), which is:
    w = SUM_i ( SVcoeff_i * SV_i )
and considering those dimensions with the highest absolute value.

NOTE: This only applies to linear SVM models, which are standard in PICA.
DO NOT use for other kernels like RBF etc.
See: http://jmlr.org/proceedings/papers/v3/chang08a/chang08a.pdf

@author: Roman V. Feldbauer
@date: 2015-02-03
"""


class SVMmodelError(Exception):
    def __init__(self, value):
        self.value = value
    def __str__(self):
        return repr(self.value)


def readMetadata(model):
    metadata = {}
    processedMetadata = 0
    for line in model:
        if line.startswith('svm_type') and not metadata.has_key('svm_type'):
            metadata['svm_type'] = line.split()[-1]
            processedMetadata += 1
```

```
elif line.startswith('kernel_type') and not metadata.has_key('kernel_type'):
    metadata['kernel_type'] = line.split()[-1]
    processedMetadata += 1
elif line.startswith('nr_class') and not metadata.has_key('nr_class'):
    metadata['nr_class'] = int( line.split()[-1] )
    processedMetadata += 1
elif line.startswith('total_sv') and not metadata.has_key('total_sv'):
    metadata['total_sv'] = int( line.split()[-1] )
    processedMetadata += 1
elif line.startswith('rho') and not metadata.has_key('rho'):
    metadata['rho'] = float( line.split()[-1] )
    processedMetadata += 1
elif line.startswith('label') and not metadata.has_key('label'):
    labels = line.split()
    if len(labels) > 3:
        raise SVMmodelError('Not binary classification')
    elif len(labels)-1 != metadata['nr_class']:
        raise SVMmodelError('nr_class and actual number of classes do not match')
    else:
        del labels[0]
        metadata['label'] = [int(label) for label in labels]
    processedMetadata += 1
elif line.startswith('nr_sv') and not metadata.has_key('nr_sv'):
    numberOfSVs = line.split()
    if len(numberOfSVs)-1 != metadata['nr_class']:
        raise SVMmodelError('nr_class does not match the number of entries in
            nr_sv')
    else:
        del numberOfSVs[0]
        metadata['nr_sv'] = [int(number) for number in numberOfSVs]
    if sum(metadata['nr_sv']) != metadata['total_sv']:
        raise SVMmodelError('Numbers of support vectors per class do not add up
            to total_sv')
    processedMetadata += 1
elif line.startswith('SV'):
    # do nothing, this line only indicates the start of SV block
    processedMetadata += 1
else:
    if processedMetadata != 8: # expecting metadata in lines 0..7, SVs from line
        8
        raise SVMmodelError('Meta data error')
    else:
        #everything alright!
        pass
#for debugging purposes
```

```python
#print metadata


    return metadata


def readSupportVectors(model, metadata):
    # MAGIC number 8: Lines of metadata in an SVM model
    # TODO remove magic number
    numberOfFeatures = len(model[8].split()) - 1
    numberOfSVs = len(model) - 8
    assert numberOfSVs == metadata['total_sv'], \
        "The number of support vectors according to metadata does not " + \
        "match the number that is present in the actual data set."

    # TODO immediately create numpy array instead of general list
    sv = [[0 for svector in range(numberOfSVs)] for feature in range(numberOfFeatures)]
    svCoeff = [0.0 for svector in range(numberOfSVs)]

    for line in xrange(8, len(model)):
        currentSupportVector = model[line].split()
        svCoeff[line-8] = float( currentSupportVector[0] )
        del currentSupportVector[0]
        if len(currentSupportVector) != numberOfFeatures:
            raise SVMmodelError('support vectors have different dimensionality')

        for dataPoint in currentSupportVector:
            presenceValue = int( dataPoint.split(':')[-1] )

            if presenceValue == 1:
                feature = int( dataPoint.split(':')[0] )
                sv[feature][line-8] = 1
            # no need to handle presenceValue=0, since matrix was init as zeros


    return sv, svCoeff

def calculateWeightsVector(sv, svCoeff):
    assert len(sv[0]) == len(svCoeff), \
        "SV matrix and svCoeff vector dimensionality do not match.\n" + \
        "SV matrix is %d x %d, svCoeff vector is %d x 1" % \
        (len(sv), len(sv[0]), len(svCoeff))
    #print "DEBUG:", str(len(sv)), str(len(sv[0])), str(len(svCoeff))

    svMatrix = numpy.array(sv)
    svCoeffVector = numpy.array(svCoeff)
    #print "DEBUG:", str(len(svMatrix)), str(len(svMatrix[0])), str(len(svCoeffVector))
    w = svMatrix.dot(svCoeffVector)
```

```
#print "DEBUG:", w


    return w


def rankDimensions(w):
    """ return a list of sorted indices of w """
    return sorted(range(len(w)), key=lambda k: abs(w[k]), reverse=True)


def readFeatureMap(featureMapFile):
    with open(featureMapFile, 'r') as handle:
        featureMap = pickle.loads(handle.read())
    return featureMap


def readClassLabelMap(classLabelMapFile):
    with open(classLabelMapFile, 'r') as handle:
        classLabelMap = pickle.loads(handle.read())
    return classLabelMap


def readNogDescription(nogDescriptionFile):
    nogDescriptionDict = {}
    with open(nogDescriptionFile, 'r') as handle:
        lines = handle.readlines()
    for line in lines:
        words = line.split('\t')
        nogDescriptionDict[words[0]] = words[1].strip()
    return nogDescriptionDict


def determinePredictionClass(w, args):
    if args.clmi:
        clmi = readClassLabelMap(args.clmi)
    elif os.path.isfile(args.model + ".classlabelmapindex"):
        clmi = readClassLabelMap(args.model + ".classlabelmapindex")
    else:
        raise SVMmodelError("Could not find class label map index file")

    if clmi[0] == 'YES' and clmi[1] == 'NO':
        pass
    elif clmi[0] == 'NO' and clmi[1] == 'YES':
        w = [-w_i if w_i != 0 else w_i for w_i in w]    #change the sign unless zero
    else:
        raise SVMmodelError("Class label map index is supposed to have values " + \
            "'YES' and 'NO', but is has %r and %r" % (clmi[0], clmi[1]) )


    return w
```

```python
def printFeatureRanking(w, dimRank, args):
    if args.fmi:
        fmi = readFeatureMap(args.fmi)
    elif os.path.isfile(args.model + ".featuremapindex"):
        fmi = readFeatureMap(args.model + ".featuremapindex")
    else:
        raise SVMmodelError("Could_not_find_feature_map_index_file")

    descriptionHeader = ''
    description = ''
    if args.descr:
        nogDescription = readNogDescription(args.descr)
        descriptionHeader = '\tGroup_description'

    absLastRank = 1.0
    relevanceThreshold = abs(w[dimRank[0]]) * (1 - args.range/100.0 )
    print "Group_ID\tScore\tClass"+descriptionHeader
    for rank in dimRank:
        assert abs(w[rank]) <= absLastRank, "Feature_ranking_list_appears_not_to_be_ \
            sorted._" + \
            "%r_<=_%r_evaluated_to_False" % (abs(w[rank]), absLastRank)
        if abs(w[rank]) >= relevanceThreshold:
            if w[rank] >= 0:
                predictorForClass = 'YES'
            else:
                predictorForClass = 'NO'
            # if fmi[rank].find('/') != -1:    #Several COGs/NOGs might be grouped
            #     together because of same profile
            featureGroup = fmi[rank].split('/')  # ... need to be split on FS '/'
            for feature in featureGroup:          # ... and printed individually.
                if args.descr:
                    description = '\t' + nogDescription[feature]
                print feature + '\t' + str(w[rank]) + '\t' + predictorForClass + \
                    description
            # else:    # Single COG can be printed directly
            #     print fmi[rank] + '\t' + str(w[rank]) + '\t' + predictorForClass +
            #     description

            absLastRank = abs(w[rank])
        else:
            break

def checkArguments(args):
    if not os.path.isfile(args.model):
        print "ARGUMENT_ERROR:_SVM_model_file_does_not_exist"
```

```
            exit(1)
        if args.clmi and not os.path.isfile(args.clmi):
            print "ARGUMENT␣ERROR:␣Class␣label␣map␣index␣file␣does␣not␣exist"
            exit(1)
        if args.fmi and not os.path.isfile(args.fmi):
            print "ARGUMENT␣ERROR:␣Feature␣map␣index␣file␣does␣not␣exist"
            exit(1)
        if args.range < 0 or args.range > 100:
            print "ARGUMENT␣ERROR:␣Range␣must␣be␣an␣integer␣in␣[0,␣100]"
            exit(1)
        if args.descr and not os.path.isfile(args.descr):
            print "ARGUMENT␣ERROR:␣NOG␣description␣file␣could␣not␣be␣found"
            exit(1)


#######################################################
#
#           MAIN PROGRAM
#
#######################################################
#
# outline
# 1. Read SVM model
# 2. Save SVs and SV_coeffs in numpy vectors
# 3. Calculate w
# 4. Select top ranking dimensions
# 5. Read classlabelmapindex and featuremapindex
# 6. Map dimensions to features and determine whether pos/neg phenotype predictor
# 7. Output sorted list of COGs/NOGs and their scores


import argparse
import numpy
import pickle
import os


defaultRange = 100
rangeHelp="""Restrict␣output␣to␣top␣features␣only.␣The␣highest␣ranking␣feature␣is␣
    returned
as␣well␣as␣those␣features␣with␣scores␣that␣are␣<=␣N␣percent␣lower
[N=0␣...␣only␣top␣feature,␣N=100␣...␣complete␣feature␣list,␣default:␣N=%d]""" %
    defaultRange
parser = argparse.ArgumentParser(version="SVM␣feature␣ranking")
parser.add_argument("model", action="store", help="SVM␣model␣FILE", metavar="FILE")
parser.add_argument("-r","--range",action="store",dest="range", help=rangeHelp, metavar="
    N", type=int, default=defaultRange)
parser.add_argument("-c","--class_label",action="store",dest="clmi", help="Class␣label␣
```

```
    map␣index␣file␣corresponding␣to␣SVM␣model", metavar="FILE")
parser.add_argument("-f","--feature_map",action="store",dest="fmi", help="Feature␣map␣
    index␣file␣corresponding␣to␣SVM␣model", metavar="FILE")
parser.add_argument("-d","--NOG_description",action="store",dest="descr", metavar="FILE",
    help="Read␣NOG␣descriptions␣from␣FILE␣and␣add␣them␣to␣the␣ouput")
args = parser.parse_args()
checkArguments(args)


with open(args.model, 'r') as modelFile:
    model = [line[:-1].strip('\n').strip('\r') for line in modelFile.readlines()]
try:
    metadata = readMetadata(model)


    sv, svCoeff = readSupportVectors(model, metadata)


    w = calculateWeightsVector(sv, svCoeff)


    dimRank = rankDimensions(w)


    w = determinePredictionClass(w, args)


    printFeatureRanking(w, dimRank, args)


except SVMmodelError as err:
    print "ERROR:␣While␣parsing␣the␣SVM␣model,␣the␣following␣error␣occurred:", err.value
```

Appendix B

Tables

Feature ranking for nitrifiers

Rank	COGID	Score	eggNOG description
1	NOG83329	0.01361	Domain of unknown function (DUF2024)
2	COG5424	0.01109	Ring cyclization and eight-electron oxidation of 3a- (2-amino-2-carboxyethyl)-4,5-dioxo-4,5,6,7,8,9-hexahydroquinoline- 7,9-dicarboxylic-acid to PQQ - (By similarity)
3	COG0348	-0.01108	4Fe-4S Ferredoxin iron-sulfur binding domain protein
4	NOG09038	0.01076	Glycosyl transferase
5	COG1838	-0.01033	fumarate
6	COG1951	-0.01025	fumarate
7	COG0155	0.01006	Component of the sulfite reductase complex that catalyzes the 6-electron reduction of sulfite to sulfide. This is one of several activities required for the biosynthesis of L- cysteine from sulfate (By similarity)
8	COG2132	0.00980	Multicopper oxidase
9	COG1861	0.00928	Acylneuraminate cytidylyltransferase
10	COG2089	0.00925	synthase
11	COG2094	0.00923	3-methyladenine DNA glycosylase
12	COG2188	-0.00908	GntR family transcriptional regulator
13	COG0646	0.00872	methionine synthase
14	COG0132	0.00863	Catalyzes a mechanistically unusual reaction, the ATP- dependent insertion of CO_2 between the N7 and N8 nitrogen atoms of 7,8-diaminopelargonic acid (DAPA) to form an ureido ring (By similarity)
15	NOG02733	0.00855	
16	COG1086	0.00855	polysaccharide biosynthesis protein
17	COG1742	0.00843	UPF0060 membrane protein
18	COG3415	0.00841	transposase
19	COG2133	0.00835	Dehydrogenase
20	COG0529	0.00833	Catalyzes the synthesis of activated sulfate (By similarity)
21	COG1404	0.00826	peptidase (S8 and S53, subtilisin, kexin, sedolisin
22	COG1216	0.00822	Glycosyl transferase, family 2
23	COG0826	-0.00816	Peptidase U32
24	COG0243	-0.00811	molybdopterin oxidoreductase

25	COG2086	-0.00811	Electron transfer flavoprotein
26	COG2116	0.00811	Formate nitrite transporter
27	COG1213	0.00810	nucleotidyl transferase
28	COG2025	-0.00810	Electron transfer flavoprotein
29	COG0839	0.00809	NADH dehydrogenase subunit j
30	COG0553	0.00809	helicase
31	COG0591	-0.00806	symporter
32	NOG14278	0.00805	BNR repeat
33	COG0785	0.00803	Cytochrome c biogenesis protein
34	NOG04041	0.00797	Nad-dependent epimerase dehydratase
35	COG2358	-0.00791	TRAP transporter, solute receptor (TAXI family
36	NOG30935	0.00790	periplasmic binding protein
37	COG1819	0.00788	glycosyltransferase
38	NOG11046	0.00782	monooxygenase, subunit B
39	COG1331	0.00781	spermatogenesis-associated protein
40	NOG92028	0.00781	alkyl hydroperoxide reductase Thiol specific antioxidant Mal allergen
41	COG3794	0.00772	Blue (Type 1) copper domain protein
42	COG0684	-0.00772	Globally modulates RNA abundance by binding to RNase E (Rne) and regulating its endonucleolytic activity. Can modulate Rne action in a substrate-dependent manner by altering the composition of the degradosome. Modulates RNA-binding and helicase activities of the degradosome (By similarity)
43	NOG11947	0.00767	dehydrogenase
44	COG1430	0.00761	exported protein
45	COG2947	0.00757	Thymocyte nuclear protein
46	COG4122	0.00755	o-methyltransferase
47	COG3484	-0.00751	20S proteasome, A and B subunits
48	COG1410	0.00749	methionine synthase
49	NOG04856	0.00747	efflux transporter, rnd family, mfp subunit
50	COG0161	0.00745	Catalyzes the transfer of the alpha-amino group from S- adenosyl-L-methionine (SAM) to 7-keto-8--aminopelargonic acid (KAPA) to form 7,8-diamino-pelargonic acid (DAPA). It is the only animotransferase known to utilize SAM as an amino donor (By similarity)

Table B.1: Top 50 most predictive features for the nitrification trait as obtained from the feature ranking algorithm.

Feature ranking for obligate intracellular microbes

Rank	COGID	Score	Group description
1	COG0129	-0,01250	Dihydroxy-acid dehydratase
2	COG0069	-0,01247	glutamate synthase
3	COG0077	-0,01235	Prephenate dehydratase
4	COG0133	-0,01219	The beta subunit is responsible for the synthesis of L-tryptophan from indole and L-serine (By similarity)
5	COG0065	-0,01197	Catalyzes the isomerization between 2-isopropyl-malate and 3-isopropylmalate, via the formation of 2-isopropylmaleate (By similarity)
6	COG0066	-0,01197	Catalyzes the isomerization between 2-isopropyl-malate and 3-isopropylmalate, via the formation of 2-isopropylmaleate (By similarity)
7	COG0134	-0,01186	indole-3-glycerol phosphate synthase
8	COG0547	-0,01186	Anthranilate phosphoribosyltransferase
9	COG0159	-0,01186	The alpha subunit is responsible for the aldol cleavage of indoleglycerol phosphate to indole and glyceraldehyde 3- phosphate (By similarity)
10	COG0119	-0,01183	Catalyzes the condensation of the acetyl group of acetyl-CoA with 3-methyl-2-oxobutanoate (2-oxoiso-valerate) to form 3- carboxy-3-hydroxy-4-methyl-pentanoate (2-isopropylmalate) (By similarity)
11	COG0028	-0,01120	acetolactate synthase
12	COG0440	-0,01101	Acetolactate synthase small subunit
13	COG0059	-0,01089	Alpha-keto-beta-hydroxylacyl reductoisomerase
14	COG0287	-0,01073	prephenate dehydrogenase
15	COG0179	-0,01071	Fumarylacetoacetate hydrolase
16	COG1052	-0,01060	Dehydrogenase
17	COG0040	-0,01059	Catalyzes the condensation of ATP and 5-phosphoribose 1- diphosphate to form N'-(5'-phosphoribosyl)-ATP (PR-ATP). Has a crucial role in the pathway because the rate of histidine biosynthesis seems to be controlled primarily by regulation of HisG enzymatic activity (By similarity)
18	COG0107	-0,01059	IGPS catalyzes the conversion of PRFAR and glutamine to IGP, AICAR and glutamate. The HisF subunit catalyzes the cyclization activity that produces IGP and AICAR from PRFAR using the ammonia provided by the HisH subunit (By similarity)
19	COG0106	-0,01059	phosphoribosylformimino-5-aminoimidazole carboxamide ribotide isomerase
20	COG0131	-0,01059	imidazoleglycerolphosphate dehydratase
21	COG0141	-0,01059	Catalyzes the sequential NAD-dependent oxidations of L- histidinol to L-histidinaldehyde and then to L-histidine (By similarity)
22	COG0118	-0,01059	IGPS catalyzes the conversion of PRFAR and glutamine to IGP, AICAR and glutamate. The hisH subunit provides the glutamine amidotransferase activity that produces the ammonia necessary to hisF for the synthesis of IGP and AICAR (By similarity)
23	COG0137	-0,01033	Citrulline–aspartate ligase
24	COG0067	-0,01027	glutamate synthase
25	COG0070	-0,01027	glutamate synthase
26	COG0111	-0,01024	Dehydrogenase

27	COG2022	-0,00992	Catalyzes the rearrangement of 1-deoxy-D-xylulose 5- phosphate (DXP) to produce the thiazole phosphate moiety of thiamine. Sulfur is provided by the thiocarboxylate moiety of the carrier protein ThiS. In vitro, sulfur can be provided by $H(2)S$ (By similarity)
28	COG0714	-0,00991	ATPase associated with various cellular activities
29	COG0139	-0,00981	Phosphoribosyl-amp cyclohydrolase
30	COG0640	-0,00972	Transcriptional regulator, arsr family
31	COG0135	-0,00964	N-(5'-phosphoribosyl)anthranilate isomerase
32	COG0352	-0,00948	Condenses 4-methyl-5- (beta-hydroxyethyl) thiazole monophosphate (THZ-P) and 2-methyl-4-amino-5-hydroxymethyl pyrimidine pyrophosphate (HMP-PP) to form thiamine monophosphate (TMP) (By similarity)
33	COG0299	-0,00941	phosphoribosylglycinamide formyltransferase
34	COG2252	-0,00938	Xanthine uracil vitamin C permease
35	COG0038	-0,00937	chloride channel
36	COG0041	-0,00931	Catalyzes the conversion of N5-carboxyaminoimidazole ribonucleotide (N5-CAIR) to 4-carboxy-5-aminoimidazole ribonucleotide (CAIR) (By similarity)
37	COG0263	-0,00930	Catalyzes the transfer of a phosphate group to glutamate to form glutamate 5-phosphate which rapidly cyclizes to 5- oxoproline (By similarity)
38	COG0014	-0,00930	Catalyzes the NADPH dependent reduction of L-gamma- glutamyl 5-phosphate into L-glutamate 5-semialdehyde and phosphate. The product spontaneously undergoes cyclization to form 1-pyrroline-5-carboxylate (By similarity)
39	COG0047	-0,00926	phosphoribosylformylglycinamidine synthase
40	COG0034	-0,00926	Glutamine phosphoribosylpyrophosphate amidotransferase
41	COG0046	-0,00926	phosphoribosylformylglycinamidine synthase
42	COG0150	-0,00926	phosphoribosylaminoimidazole synthetase
43	COG0151	-0,00926	Phosphoribosylglycinamide synthetase
44	COG0476	-0,00920	UBA THIF-type NAD FAD binding protein
45	COG0414	-0,00916	Catalyzes the condensation of pantoate with beta-alanine in an ATP-dependent reaction via a pantoyl-adenylate intermediate (By similarity)
46	COG0413	-0,00916	Catalyzes the reversible reaction in which hydroxymethyl group from 5,10-methylenetetrahydrofolate is transferred onto alpha-ketoisovalerate to form keto-pantoate (By similarity)
47	NOG00108	-0,00915	Dehydrogenase
48	COG0347	-0,00910	Nitrogen regulatory protein pii
49	COG0031	-0,00908	cysteine synthase
50	COG0351	-0,00906	phosphomethylpyrimidine kinase

Table B.2: Top 50 most predictive features for the obligate intracellular trait as obtained from the feature-ranking algorithm.

Appendix C

Acknowledgments

First and foremost, I acknowledge and cordially thank my supervisor Thomas Rattei for giving me the opportunity to work on this highly interesting topic, for his ongoing support and constructive criticism, for giving me just the right ratio of guidance and freedom in the project, and of course for all the free organic fruits.

I also acknowledge my cooperation partners Frederik Schulz, Matthias Horn, Holger Daims and Alex Tveit, who kindly provided me with data necessary for important parts of this project, and gave precious input for interpretation of the results. The cooperation with Frederik and Matthias made my first publication possible, for which I am very grateful.

Many thanks go to all my colleagues at CUBE, the Thomases, Markus, Hans-Jörg, Dmitrij, Alex, Javier, Britta, Miriam, Daniel, Stefanie, Florian, and Franziska for providing such a harmonic working atmosphere and also valuable advice.

I want to express my deepest gratitude to my family, to my parents Luise and Norbert, to whom I owe to having a childhood free from worry, who provided an always loving environment to grow up in, with just the right ratios of security and freedom as well as technics and art for my personal development, and who continue to support me in any situation. Thanks a bunch to my big brother Gregor for being such a good role model. We'll see, if I ever catch up... Finally, many thanks to Peter and Helma for all the intellectually most inspiring dinners! I am also very grateful to my fellow students Christoph, Valerie, Elisa, Agathe, Jan, Stefanie and Bernhard. Student days would have been wasted unless shared with nice and smart people like you. Let's never stop discussing life, the universe and everything!

Cheers to all my friends! Thank you, Denise, Thomas, Bernhard, Katharina, Mira and Benedikt, you are always most welcome on my ship; Viktoria, for all the shared time, wines and newspapers; Stefanie, thanks for your delightful Bavarian; Christoph and Johannes, for all the good times together as 'highwaymen'; Theresa, for your lively, open-minded, at times emphatic nature; Julia, Steffi, Ana, Thomas, Sara and all the other good people, who enrich my life. And Franziska, I wouldn't be same if it wasn't for the time with you.

Printed in the United States
By Bookmasters